The Concept of Drive in Classical German Philosophy

Manja Kisner • Jörg Noller
Editors

The Concept of Drive in Classical German Philosophy

Between Biology, Anthropology, and Metaphysics

palgrave
macmillan

Editors
Manja Kisner
University of Wuppertal
Wuppertal, Germany

Jörg Noller
University of Munich
Munich, Germany

ISBN 978-3-030-84159-1 ISBN 978-3-030-84160-7 (eBook)
https://doi.org/10.1007/978-3-030-84160-7

Cover illustration: Maram_shutterstock.com

This Palgrave Macmillan imprint is published by the registered company Springer Nature Switzerland AG.
The registered company address is: Gewerbestrasse 11, 6330 Cham, Switzerland

Acknowledgments

Most of the chapters included in this volume were presented at the conference "The Concept of Drive in Classical German Philosophy," which was held at the Ludwig-Maximilians-Universität in Munich on October 11 and 12, 2019. We would like to thank the Fritz Thyssen Stiftung for generously sponsoring the conference and the participants for presenting and discussing their papers. We would also like to thank Jakob Grüner for his editorial support with the volume.

—Manja Kisner and Jörg Noller

Contents

Notes on Contributors

Yoon H. Choi is an assistant professor in the Department of Philosophy at Marquette University, USA. Prior to that, she was a postdoc at Tufts University. She is working on an account of Kant's theory of self-consciousness and self-cognition that understands the self in question as the human being: an embodied rational agent who essentially feels her existence as much as she thinks it.

Alix Cohen is a reader at the University of Edinburgh, UK. She is the author of *Kant and the Human Sciences: Biology, Anthropology and History* (2009) and has written articles and book chapters on Kant, including most recently "A Kantian Account of Emotions as Feelings" (*Mind*) and "Kant on Epistemic Autonomy" (*Proceedings for the 13th International Kant Congress*). She is the co-editor of the *British Journal for the History of Philosophy*.

Andrew Cooper is an assistant professor in the Department of Philosophy at the University of Warwick, UK. His work is inspired by modern European philosophy, especially Kant and post-Kantian thought, and deals with a range of issues in philosophy of science, metaphysics, and aesthetics. He is author of *The Tragedy of Philosophy: Kant's Critique of Judgment and the Project of Aesthetics* (2016).

Nigel DeSouza is an associate professor in the Department of Philosophy at the University of Ottawa, Canada. He works on the philosophy of Herder, early modern philosophy, and contemporary ethics. He has written articles on Herder's metaphysics, epistemology, philosophy of language, philosophy of history, and moral philosophy, as well as on the foundations of ethical agency. His articles have appeared in *The British Journal for the History of Philosophy*, *Intellectual History Review*, *Herder Yearbook*, the *Herder Handbuch*, and *Ethical Theory and Moral Practice*. He is the co-editor of the volume *Herder: Philosophy and Anthropology* (2017).

George di Giovanni is Professor of Philosophy Emeritus at McGill University, Canada, specializing in the Late Enlightenment and in German Idealism. He has introduced Fr. H. Jacobi in English with the translation and analysis of his works (1994 & 2009); retranslated Hegel's *The Science of Logic* (2010), and contributed to the Cambridge edition of Kant (1996). His many publications, in various international venues, include *Freedom and Religion in Kant and His Immediate Successors: The Vocation of Humankind, 1774–1800* (2005) and *Hegel and the Challenge of Spinoza: A Study in German Idealism, 1801–1831* (2021).

Dietmar Heidemann is Professor of Philosophy at the University of Luxembourg, Luxembourg. His specialization is Kant and German Idealism, especially Hegel. He is the first chairman of the Kant-Gesellschaft and a member of the Kant-Kommission of the Berlin-Brandenburgischen Akademie der Wissenschaften. Among his publications are *Der Begriff des Skeptizismus: Seine systematischen Formen, die pyrrhonische Skepsis und Hegels Herausforderung* (2007) and *Kant und das Problem des metaphysischen Idealismus* (1998). He is also the editor-in-chief of the *Kant Yearbook*. His recent book is *L'idéalisme: Une introduction à la philosophie classique allemande* (foreseen for 2021).

Manja Kisner is lecturer in Philosophy at the University of Wuppertal, Germany. She completed her Ph.D. and postdoc studies from the University of Munich. Her research focuses on classical German philosophy and Schopenhauer. She is the author of *Der Wille und das Ding an sich: Schopenhauers Willensmetaphysik in ihrem Bezug zu Kants kritischer Philosophie und dem nachkantischen Idealismus* (2016) and has written articles and book chapters on Kant, Fichte, Schelling, and Schopenhauer.

Jörg Noller is lecturer in Philosophy at the University of Munich, Germany. He is the author and editor of a number of books, editions, and articles on freedom from a historical and systematic perspective, including *Die Bestimmung des Willens* (2016), *Kant's Early Critics on Freedom of the Will* (with John Walsh, 2021), "Kant and Reinhold's Dilemma" (*European Journal of Philosophy*, 2019), and "Schiller on Freedom of the Will" (*European Journal of Philosophy*, 2020). His research interests include metaphysics, freedom, personhood, and German Idealism.

Angelica Nuzzo is Professor of Philosophy at the Graduate Center and Brooklyn College, City University of New York (CUNY), USA. Among her books are *Approaching Hegel's Logic, Obliquely: Melville, Molière, Beckett* (2018), *Memory, History, Justice in Hegel* (2012), and *Ideal Embodiment: Kant's Theory of Sensibility* (2008).

Marco Segala is Professor of History of Philosophy at the University of L'Aquila, Italy. He graduated from the Scuola Normale Superiore of Pisa (1987) and obtained his Ph.D. in Philosophy from the University of Firenze (1992). Former "Alexander von Humboldt" fellow and "Marie Curie" fellow, his research activity spans history of post-Kantian philosophy—focusing on Schopenhauer, history of science (1750–1950), and history and philosophy of music (1800–1900). He is active in the digital humanities as coordinator of the Ampère digital project at the Centre Alexandre Koyré, Paris, and the Schopenhauer digital project at the Schopenhauer-Archiv, Frankfurt am Main.

Violetta L. Waibel is Professor of European Philosophy at the University of Vienna, Austria. Her research focuses on theoretical philosophy (conscience, subjectivity, inter-subjectivity, and space and time), the relation of cognition and emotion, as well as aesthetics in Kant, German Idealism, Romanticism, and Modernity. She was the organizer of the 30th International Hegel-Congress in 2014 and of the 12th International Kant-Congress in 2015. Her key publications include *Hölderlin und Fichte: 1794–1800* (2000), *Affektenlehre und amor Dei intellectualis. Die Rezeption Spinozas im Deutschen Idealismus, in der Frühromantik und in der Gegenwart* (ed., 2012), and *Fichte und Sartre über Freiheit: Das Ich und der Andere* (2015).

Catherine Wilson FRSC is based in Berlin, Germany. She is also from time to time a Visiting Presidential Professor at the City University of New York Graduate Center, USA. She has written widely on seventeenth- and eighteenth-century science and metaphysics, most recently on the history and theory of atomism and materialism. Her books include *Leibniz's Metaphysics* (1989/2015), *The Invisible World* (1995/2009), and *Epicureanism at the Origins of Modernity* (2008). Her study of Kant: *Kant and the Naturalistic Turn: Futility and Transcendence* is under contract from Oxford University Press to appear in 2022.

John H. Zammito is Baker College Chair for History of Science, Technology, and Innovation in the Department of History at Rice University, USA. His research focuses on Kant and his student and rival Herder. Key publications include *The Genesis of Kant's Critique of Judgment* (1992), *Kant, Herder, and the Birth of Anthropology* (2002), and *The Gestation of German Biology: Philosophy and Physiology from Stahl to Schelling* (2018).

Paul Ziche is Professor (chair) of History of Modern Philosophy at Utrecht University, Netherlands. He completed his Ph.D. and "Habilitation" from the Ludwig-Maximilians-University of Munich, and his earlier positions were at the Institute for the History of Medicine, Science, and Technology, Friedrich Schiller University, Jena, and the Bavarian Academy of Sciences and Humanities, Munich. His areas of research include German Idealism, with a focus upon Schelling and topics in the philosophy of nature; the interaction between philosophy and the sciences, mainly around 1800 and around 1900; new forms of "science" and philosophy in the nineteenth century; and history of disciplines and (scientific) institutions.

Günter Zöller is Professor of Philosophy at the University of Munich, Germany. He studied at the University of Bonn, the École normale supérieure, Paris, and Brown University. He has held visiting professorships at Princeton University, Emory University, McGill University, Seoul National University, The Chinese University of Hong Kong, and the University of Bologna. His more recent book publications include *Kant,*

Anthropology, History, and Education (2008), *Der Staat als Mittel zum Zweck* (Nomos, 2011), *Fichte Lesen* (2013), *Res Publica* (2015), *The Cambridge Companion to Fichte* (2016), *Philosophie des 19. Jahrhunderts* (2018), and *Hegels Philosophie: Eine Einführung* (2020).

1

Introduction

Manja Kisner and Jörg Noller

Philosophy has generally pursued dialogue with other disciplines, and in particular with the natural sciences. The period of classical German philosophy is no exception. In this period, Kant was impressed by the great success of physics, especially Newtonian mechanics in the seventeenth century. In the first *Critique,* he used physics as a model for his own critical project, which aimed at turning philosophy into a proper science. But it is not only physics that the German philosophers at the end of the eighteenth century engaged with;the emerging life sciences were of great importance for them as well. Although "biology" as the name for a discipline with a distinct research program first appeared in the works of Gottfried Reinhold Treviranus and Jean-Baptiste Lamarck in 1802, it was

M. Kisner (✉)
University of Wuppertal, Wuppertal, Germany
e-mail: kisner@uni-wuppertal.de

J. Noller
University of Munich, Munich, Germany
e-mail: joerg.noller@lrz.uni-muenchen.de

M. Kisner, J. Noller (eds.), *The Concept of Drive in Classical German Philosophy*,
https://doi.org/10.1007/978-3-030-84160-7_1

the preceding physiological and medical research in the eighteenth century that paved the way for biology as an independent discipline.[1]

Taking into account this pivotal time in history, in which biology as a scientific disciple emerged, it is not at all surprising that philosophers were well acquainted with the empirical research that shook the commonly held views about life and organisms. Philosophers discussed from their perspectives the special status of organic nature and engaged with the pressing question of whether a science of organic nature, which would differ substantially from the mechanical and physical sciences, is possible at all. In the third *Critique*, Kant discussed the possibility of a special science of organisms, and in doing so laid the foundations for a prospective philosophy of biology. After Kant, the German idealists would continue the debate initiated in the third *Critique*, and expand the scope of philosophy of nature to the issues of biology. Since the 1980s, we can observe a growing scholarly interest in these topics and in the mutual impact that philosophy and emerging biology had on each other. For the study of classical German philosophy in particular, however, it is the encounter with German physiologists and physicians at the universities of Göttingen and Halle, such as Stahl, von Haller, and Blumenbach, that is of special interest.

Timothy Lenoir's seminal work *The Strategy of Life: Teleology and Mechanics in Nineteenth Century German Biology* (1982) goes so far as to claim that "the development of biology in Germany during the first half of the nineteenth century was guided by a core of ideas and a program for research set forth initially during the 1790s" (Lenoir 1982, 2). According to his view, it is Kant who significantly influenced this development. What Lenoir recognizes as a special merit of Kant's approach is favoring the mechanistic approach of sciences over the teleological views of organic nature, which we cannot totally resist, but which also cannot be used in a proper scientific program. Lenoir claims that the development of

[1] Peter McLaughlin (2002) notes, however, that the word "biology" has a longer history that goes back to Michael Christoph Hanov, who used the word in 1766 in the title of his book *Geology, Biology, General Phytology and Dendrology, or the Science of the Earth, of Living Things and of Vegetating Things in General* (*Philosophiae naturalis sive physicae dogmaticae: Geologia, biologia, phytologia generalis et dendrologia*). But Hanov did not yet use the term "biology" to describe an independent scientific discipline.

biological sciences in the nineteenth century was possible because those researchers followed the Newtonian framework and rejected teleological explanations of organic nature. According to Lenoir, the life scientists of the nineteenth century agreed with Kant that the mechanical explanations are the only appropriate ones for a proper science, but they diverged from Kant's view that there will never be a Newton of the blade of grass and, consequently, that a separate science of organic nature, besides physics, is not possible.

Despite the important role that Lenoir ascribes to Kant, he is at the same time highly skeptical about the influence of German *Naturphilosophie* on the emerging discipline of biology. Unlike Kant, the *Naturphilosophie* fearlessly embraced a teleological worldview and assigned it a constitutive role in the study of organisms and nature as a whole. The teleology that post-Kantian idealists as well as the Romantics had in mind was not an *external* teleology based on the idea of the intelligent design, but instead an *internal* teleology that describes nature and organisms as self-propagating and self-organizing processes, which cannot be properly grasped only by reference to mechanical explanations. According to Lenoir, this approach is not appropriate for a scientific program that seeks to establish biology as an independent scientific discipline with its own agenda. In contrast, he views the progress of biology as a "result of the elimination [of] rather than a concerted commitment to teleological thinking" (Lenoir 1982, 8).

This highly influential interpretation by Lenoir was called into question by later historians of science and philosophy, such as Beiser (2002) or Richards (2002) and, most recently, by Zammito (2018) and Gambarotto (2018). In *The Gestation of German Biology*, Zammito reconstructs the development of philosophy and physiology from Stahl to Schelling and thereby points to a crucial role that the *Naturphilosophie* and especially Schelling's philosophy of nature had on the emergent science of biology. In a similar vein, Gambarotto argues in his work *Vital Forces, Teleology, and Organization* "that Romantic *Naturphilosophie* played an important role in the rise of biology in Germany during this period" and that this was, above all, due to how post-Kantian philosophers thought about teleology, namely by understanding internal teleology and purposive organization as constitutive for the rise of life sciences (Gambarotto 2018, xv).

Due to the research done by physiologists and physicians, the conviction that mechanical natural philosophy cannot sufficiently describe living beings became pervasive in the second half of the eighteenth century, and life scientists began to search for the non-mechanical forces constitutive of organic nature. Blumenbach's idea of a special formative power, which he called a "formative drive" (*Bildungstrieb*), plays a crucial role in this development and also proves highly important for Kant's philosophical appropriation of the concept of drive (*Trieb*) as well as for later post-Kantian drive theories. At the end of the eighteenth century, the concept of drive started to play a central role in classical German philosophy, not only for the foundation of theoretical philosophy but also for practical philosophy. At that time, the concept of drive had evolved into a notion that could be used to describe the special characteristics of organic nature more generally; at the same time, the concept could also be employed more concretely for describing the dynamics of human life, which shapes and forms itself in freedom, yet always has natural roots. In this second sense, the philosophical concept of drive makes it possible to think of human beings not as merely naturally describable but also as cultural and moral beings who possess freedom and who develop and evolve personally, historically, and culturally.

The fourteen chapters collected in this volume investigate the philosophical history of the concept of drive in classical German philosophy and try to trace various sources of this concept prior to classical German philosophy, including biological, psychological, and anthropological accounts found in Leibniz's philosophy and the philosophy of German early Enlightenment. The volume is based on the conviction that the concept of drive is significant for the history of ideas in that it makes it possible to systematically connect different disciplines and questions and to bring them into a fruitful discussion. To this end, the volume covers a wide array of drive theories in order to offer a comprehensive historical overview of the concept of drive and discuss it both historically and systematically.

Chapter 2, by *Catherine Wilson*, sets the scene by examining the influence of Leibniz on the later drive theories in classical German philosophy.

Wilson exposes Leibniz's philosophical and conceptual contributions to the theories of Kant, Herder, and Schelling. She thereby focuses on Leibniz's vis viva controversy, on his notion of active force and on the drive of perfection. But as Wilson emphasizes, Leibniz's conception of active force had no role to play in animal activity, since this force is still conceived as mechanical force and has to be clearly distinguished from the later concept of drive that was used to describe non-mechanical formative powers. As a result, Wilson calls into question the direct influence of Leibniz on *Naturphilosophie,* and points to other sources, such as Reimarus' theory of animal drives, that were of great importance for the drive theories at the end of the eighteenth century as well.

This theme is continued in Chap. 3, in which *John H. Zammito* draws upon Reimarus' pioneering work on animal drives and presents it as a decisive source for Blumenbach's conception of *Bildungstrieb*. Zammito argues that Blumenbach's *Bildungstrieb* cannot be identified with other living and formative forces, since *Bildungstrieb* presents a special formative power unique to organic nature. Blumenbach's distinction between *Bildungstrieb* and *Bildungskräfte* was, moreover, crucial to Kant, who explicitly mentions Blumenbach in his third *Critique.* In the last part of the chapter, Zammito critically reflects on this very productive interaction between Blumenbach and Kant.

The following three chapters present different accounts of drives in the philosophies of Herder, Goethe, and Jacobi since the 1760s, which were important for the later romantic and idealistic drive theories. First, Chap. 4 by *Nigel DeSouza* offers an introduction into Herder's philosophy of life, which proves important for Kant's dispute with Herder as well as for a more enthusiastic reception of Herder in the *Naturphilosophie* and by the German idealists. For this purpose, DeSouza studies Herder's concept of force (*Kraft*), which he interprets both as foundational for his philosophy of life and for his account of drives that emerge out of forces. He explores the origins of Herder's concept of force in the philosophies of Stoics, Leibniz, Baumgarten, and Shaftesbury, and interprets Herder's philosophy not as a substance-based ontology but as a particular force-based ontology.

Chapter 5 by *Andrew Cooper* presents Goethe's little-known drive theory that he developed in his morphological writings. Cooper shows that

Goethe's notion of the "economy of nature" replaces the Newtonian model of force with an experimental conception of the formative drive. Drawing on the programs of comparative anatomy outlined by Johann Friedrich Blumenbach and Carl Friedrich Kielmeyer, Goethe defines the formative drive as a distributive principle that helps to explain structural features of organic systems. In this way, Goethe's formative drive is on the one hand constrained to the structural plan of organic systems, but on the other hand it can freely allocate the resources that organisms use and thus gives rise to new shapes. By way of conclusion, Cooper argues that the role of Goethe's morphological writings in the formation of evolutionary science is greater than we might expect at first glance.

Chapter 6 by *George di Giovanni* examines Jacobi's conception of drive as opposed to Lessing, Mendelssohn, and Kant. Jacobi's conception of drive becomes clear against the background of his conception of reason, which he understands as a form of life. The concept of drives plays a central role as a bodily and organically grounded motivating power that lets us seek the truth. As such, it serves as a complex philosophical element in Jacobi's philosophy that integrates both reason and sensibility.

The next three chapters are dedicated to Kant's concept of drive and his notion of life. In Chap. 7, *Manja Kisner* compares Kant's concept of *Triebfeder* in the *Groundwork of the Metaphysics of Morals* and in the second *Critique* with the concept of *Trieb* as Kant uses it in the second part of the third *Critique*. Kisner claims that whereas Kant's practical *Triebfeder* is still closely bound to a mechanical worldview, his concept of *Trieb* has a completely different conceptual background and is related to Kant's principle of internal purposiveness, which in the third *Critique* has only a heuristic status. The difference between *Triebfeder* and *Trieb* thus brings to the fore an important shift of focus within Kant's philosophy.

In Chap. 8, *Dietmar Heidemann* discusses the concept of drive in the third *Critique* from a broader perspective, which is not only limited to Kant's teleology but also takes into account his philosophy of history and culture. For this purpose, Heidemann focuses on Kant's idea of the drive to society, which already played an important role prior to the third *Critique*, namely in the *Idea for a Universal History*. Moreover, in this chapter Heidemann also points to the seventeenth- and eighteenth-century debate of "drives" in moral and psychology textbooks by authors such as Thomasius and Crusius, who offer important historical background for Kant's drive theory.

The last chapter of this triad explores Kant's notion of the "feeling of life" and compares it with his idea of vital powers. In Chap. 9, *Yoon H. Choi* and *Alix Cohen* focus on Kant's theory of the beautiful and the sublime in the third *Critique* in order to distinguish between Kant's feeling of the promotion of life and the feeling of the sublime, which is related to the inhibition of the vital powers. In the feeling of the promotion of life we are aware of ourselves as a part of nature and as embodied human beings. In the feeling of the sublime, by contrast, we experience ourselves as pure rational beings and become aware of our supersensible nature.

Shifting the focus to the developments after Kant, the rest of the volume deals with the idealistic and romantic conceptions of drives and concludes with Schopenhauer as a student of Blumenbach. Chapter 10 by *Günter Zöller* situates Fichte's conception of drive in the history of philosophy and distinguishes it from Plato's as well as Kant's accounts of self-rule. Whereas Plato described the functioning of the soul in analogy with the team of horses and the charioteer that has to handle and coordinate them, Kantian self-rule turns into an autonomous self-legislation. In contrast to both Plato and Kant, Fichte's comprehensive theory of the mind depends completely on drives that motivate human actions and give them orientation; as Zöller argues, in this account not much is left to the driver but all the more to the drives that govern our actions.

The next two chapters are dedicated to romantic and aesthetic conceptions of drives. *Violetta L. Waibel* in Chap. 11 analyzes the notion of drive in Hardenberg/Novalis and thereby focuses on the crucial relationship between reason and the unconscious as well as between thinking and feeling in his *Fichte Studies*. She shows how Novalis' concept of drive plays an important role in connection with a human being's mental actions and traces back its Kantian and Fichtean origins.

Jörg Noller in Chap. 12 shows how Reinhold and Schiller, following Kant's distinction between drive (*Trieb*) and driving-force (*Triebfeder*), attempted to develop a unified account that integrates both into a conception of freedom of the will. Whereas Reinhold argues for a unity between the selfish and the unselfish drive, Schiller argues for a harmony model, according to which the form-drive and the material drive are sublated in the concept of the play-drive.

Chapter 13 by *Paul Ziche* offers a comprehensive account of Schelling's notion of drive, which plays a crucial role in his understanding of sciences as well as in his theory of cognition. Ziche argues that with the concept of drive Schelling tries to overcome the traditional distinction between lower and higher epistemic capacities. In Schelling's philosophy of nature, the concept of drive is closely related to the term "instinct" and describes immediate and involuntary actions, but at the same time drives can also acquire higher cognitive functions and hence give a framework for understanding our cognition and the dynamics of sciences.

Chapter 14 by *Angelica Nuzzo* argues that in Hegel's mature philosophy the concept of drive is not exclusively restricted to his philosophy of nature, but rather plays an important role in his philosophy as a whole. Although the concept of drive is not a *terminus technicus* for Hegel, it is of systematic importance to his views on the dialectical process that bridges the gap between nature and freedom. To show this, Nuzzo focuses on Hegel's *Logic* and analyzes the dialectical and methodological function of the concept.

The final chapter (Chap. 15) by *Marco Segala* interprets Schopenhauer's notion of drive in light of his metaphysical conception of the will. For this purpose, Segala analyzes the relationship of Schopenhauer's account to those conceptions of the eighteenth century that can be found in Reimarus, Blumenbach, and Goethe, and also analyzes the systematic role that the concept of drive plays in Schopenhauer's philosophical system. Moreover, Schopenhauer's conception of drive not only stands in the philosophical tradition of classical German philosophy but also points ahead to the specific modern notion of drive as Sigmund Freud used it.

The editors hope that this volume will shed new light on a hitherto understudied subject in classical German philosophy, and will be not only of historical but also of systematic interest for those interested in the connection between classical German philosophy and the life sciences.

References

Beiser, Frederick. 2002. *German Idealism: The Struggle against Subjectivism, 1781–1801*. Cambridge: Harvard University Press.

Gambarotto, Andrea. 2018. *Vital Forces, Teleology, and Organization: Philosophy of Nature and the Rise of Biology in Germany*. Cham: Springer.

Lenoir, Timothy. 1982. *The Strategy Of Life: Teleology and Mechanics In Nineteenth Century German Biology*. Dordrecht: Reidel Publishing Company.

McLaughlin, Peter. 2002. Naming Biology. *Journal of the History of Biology* 35: 1–4.

Richards, Robert. 2002. *The Romantic Conception of Life: Science and Philosophy in the Age of Goethe*. Chicago: University of Chicago Press.

Zammito, John. 2018. *The Gestation of German Biology: Philosophy and Physiology from Stahl to Schelling*. Chicago: The University of Chicago Press.

2

The Theory of Drive: The Dual Legacy of Leibniz's Theory of Appetition

Catherine Wilson

In the late eighteenth and early nineteenth centuries, will and drive—*Wille* and *Trieb*—were held by German philosophers to be manifested throughout nature and human society. These concepts, along with references to internal forces and impulses, had appeared in contexts as varied as Kant's non-moral *Triebe* for "food, sex, rest, and movement, and […] for honour, for enlarging our cognition" (MS 6:215); Blumenbach's *Bildungstrieb* for embryology; Schiller's sensuous "*Stofftrieb*"; his Apollonian "*Formtrieb*" and aesthetic "*Spieltrieb*," and the desire-driven evolutionary theory of Lamarck. Schopenhauer emphasized the sex drive, and in the late nineteenth and early twentieth centuries, drives entered psychology as universal human needs, demanding satisfaction or causing neurosis on account of their repression, with Freud even introducing a "*Todestrieb*" opposed to every value of survival, pleasure, and creative construction.

In his 1925 study of theories of instinct, E. C. Wilm proposed Leibniz as a source of ideas fruitful for nineteenth-century German

C. Wilson (✉)
City University of New York Graduate Center, New York, NY, USA

M. Kisner, J. Noller (eds.), *The Concept of Drive in Classical German Philosophy*,
https://doi.org/10.1007/978-3-030-84160-7_2

Naturphilosophie, in which volitional concepts, including *Trieb*, play a central role. Wilm saw Leibniz as initiating a tradition of what he called "voluntarism," whose "central conception was that of an immaterial principle, analogous to will, working with more or less conscious design to reach rational ends."[1] It was this notion, he says, "that was taken up and given a fruitful application by a whole school of writers from Fichte to Bergson."[2] Frederick Beiser echoed this view. The Cartesian conception of matter as inert extension left unacceptable explanatory gaps:

> Since neither mind nor life are conceivable in spatial or mathematical terms, this made it impossible to explain them according to the laws of nature. As long as this concept of matter prevailed, there could be only those two unsatisfactory options in the philosophy of mind: dualism or materialism. The only escape from these extremes, the romantics believed, lay in going back to a competing concept of matter [...] as living force, *vis viva*. It was this concept that Leibniz had once cited against Cartesianism and that the *Naturphilosophen* now intended to revive.[3]

Rather than taking matter as passive, *Naturphilosophie* unified living and nonliving nature as a realm of active forces whose teleologies were her own, not assigned and impressed by a supernatural divinity. As Keith Peterson explains it, *Naturphilosophie* "focused on dynamical, organic, synthetic, and holistic accounts of the natural world, integrating human beings into that world rather than severing them from it."[4] Unlike Kant's Transcendental Philosophy, *Naturphilosophie* supposed that "the real necessary conditions of objects are seen to be engendered through essential forces in the material world. 'Forces' are the empirical manifestation of nature's 'productivity' or activity, and all matter, organic or inorganic, is engaged in play of forces, both free and constrained."[5]

Wilm routed the purported evolution of Leibnizianism through Hermann Samuel Reimarus, the first student of natural history to devote

[1] Wilm (1925, 123).
[2] Wilm (1925, 123).
[3] Beiser (2000, 32–33).
[4] Peterson (2004, xi).
[5] Peterson (2004, xvii).

a book to animal drives in 1760.[6] Yet few philosophers at this time, Reimarus included, associated Leibniz with a behavioural or psychologistic theory of drive or with a vision of the universe as pulsating with activity. Leibniz's monadology was regarded by those who were familiar with it as a metaphysical fantasy.[7] His philosophy was not clearly distinguished from that of his follower Christian Wolff until later in the eighteenth century, and he was known chiefly on the basis of his *Theodicy*, his theory of pre-established harmony, and the *vis viva*, which, despite its name, had no role to play in the animal activity or physiology. The deterministic *Theodicy* implied that God had arranged the whole course of nature down to its least detail, and his theory of pre-established harmony between body and soul, implied that the soul had no power to initiate changes in the body. Unknown were the *New Essays* with their extensive discussion of unconscious perception and volition, first published in 1765.[8]

Leibniz himself would have rejected *Naturphilosophie* as occult had he lived long enough to encounter it, and Reimarus did not see Leibniz as a forerunner of his views. So Wilm's thesis along with other claims for a Leibnizian influence on later theories of drive needs re-examination. In this chapter, I explore how one *oeuvre*—different parts of which came to light at different times—could give rise to such disparate, indeed opposed interpretations. I review first the features of the Leibnizian corpus that were favourable to nineteenth-century *Naturphilosophie*, and second, the features of Leibniz's philosophy that were inhospitable and that explain Reimarus's aversion. These different elements are for the most part well understood; it is their relative importance in the later context that is controversial.

[6] Reimarus (1760). On Reimarus's animal psychology and influence, see Jaynes and Woodward (1974a, b), Cheung and Cheungs (2006), Ortlieb (2010), Zammito (2017).

[7] The first German translation appeared in 1720. The metaphysic of monads was criticized by, among others, Langhansen (1721) and Müller (1745). It was ridiculed by Voltaire and Kant.

[8] On Leibniz's publication history in Germany, see Tonelli (1974).

1 Leibniz's Contributions to the Theory of Drive

Leibniz's contributions can be summed up under four subheadings:

1. "The Restoration of Activity to the Concept of Substance"
2. "The Living Individual with Perception and Appetite as the 'Atom': The Fundamental Element of Nature"
3. "The Tendency to Perfection"
4. "The Theory of Indistinct Knowledge, the Unconscious, and Its Motivating Force and Creative Power: Innateness"

1.1 The Restoration of Activity to the Concept of Substance

In his paper "On the Correction of Metaphysics and the Concept of Substance," anticipating his *Specimen Dynamicum* appearing in the *Acta Eruditorum* of 1694, Leibniz declared that "the concept of *forces* or *powers*, which the Germans call *Kraft* and the French *la force* [...] brings the strongest light to bear on our understanding of the true concept of *substance*."[9]

Spinoza, whose metaphysics Leibniz encountered early in his career, was his first target, along with Gassendi, whose Epicurean atomism had attracted him as a teenager. Spinoza claimed that individual persons, objects, and animals were only "modes" of a single divine substance endowed with will and understanding; for Leibniz individual living beings were as real as God. Gassendi favoured the corporeal atom as the fundamental unit of nature, allowing even for a corporeal soul; for Leibniz, matter was a phenomenon dependent on incorporeal, unextended soul-like atoms that were the fundamental units.

Leibniz backed his "corrections" with an appeal to physics. By 1684, he had established to his satisfaction that "motive force" measured by

[9] Leibniz, *On the Correction of Metaphysics* cited from Leibniz (1969, 433, cf. 1875–1890, IV: 469 for the untranslated text).

mv^2, which he termed the *vis viva*, as distinct from Descartes' "quantity of motion" *mv*, was conserved in all interactions, a result he presented in his "Brief Demonstration," published in the *Acta Eruditorum*. In his *Discourse on Metaphysics*, published two years later, he asserted that "[t]he distinction between force and quantity of motion is important [...] to show that we must have recourse to metaphysical considerations apart from extension in order to explain the phenomena of bodies."[10]

Leibniz now had the "incorporeal principle" he had demanded almost from the beginning of his investigations into the foundations of physics to hand. The "crude notion" of corporeal substance, Descartes' notion of matter as pure extension, seems to imply that material objects cannot resist or reflect motion, or do work, and it is a miracle how solid objects hold together. The occasionalists—and in some passages Descartes himself—addressed this problem by recourse to the mediaeval doctrine of occasionalism: God is the only true power or casual force in the universe. In the same work, Leibniz claimed to have rehabilitated the scholastic notion of a "substantial form," and he adhered to a version of hylomorphism for the remainder of his life. Bodies, as he expressed it in a late dialogue completed in 1715, "are composed of two natures—an active primitive force [...] and matter or passive primitive force", the latter inducing resistance.[11]

In his paper *De Ipsa Natura* of 1698, attacking the occasionalist J.C. Sturm, Leibniz insisted:

> If [...] the law set up by God does in fact leave some vestige of him expressed in things [...] then it must be granted that there is a certain efficacy residing in things, a form or force such as we usually designate by the name of nature, from which the series of phenomena follows according to the prescription of the first command.[12]

[10] Leibniz, *Discourse on Metaphysics* cited from Leibniz (1969, 315, cf. 1875–1890, IV: 441 for the untranslated text).

[11] Leibniz, *Conversation of Philarète and Ariste* cited from Leibniz (1989, 264, cf. 1875–1890, VI: 588 for the untranslated text).

[12] Leibniz, *On Nature Itself*, cited from Leibniz (1969, 501, cf. 1875–1890, IV: 507 for the untranslated text).

Dynamics was, as Leibniz put it, the study of "living forces"—forces that are "joined with actual motion." In the *Specimen Dynamicum* of 1695, he termed the forces systematized by mathematical physics and manifested in observable interactions, "derivative" as opposed to "primitive." Derivative active force is the power of one body to move another. Derivative passive force is the power of a body to resist being moved, "laziness or repugnance to motion," and its tendency to remain as it is.

Primitive force pertained to the "monads," the incorporeal atoms, which, like the human soul, were endowed with two powers: perception and appetition.[13] "The nature of substance necessarily requires and essentially [envelopes] progress or change, without which it would not have the force to act."[14] A monad—a human, animal, or plant soul, or simply an incorporeal atom—undergoes a predetermined series of perceptual states that are not caused by external stimuli but that flow from its concept. The passage from one perceptual state to another is appetition: though preprogrammed by God, the monad has an urge to experience next what it is determined to experience. "[P]rimitive forces can be nothing but the internal strivings [*tendentia*] of simple substances, strivings by means of which they pass from perception to perception in accordance with a certain law of their nature, and at the same time harmonize with one another, representing the universe in different ways."[15] So appetitive forces that expressed themselves derivatively in the interactions studied in mathematical physics were felt primitively and subjectively in appetite, desire, and impulse in humans, other living beings, and monads.

[13] Leibniz, *Monadology* §§ 14–19 cited from Leibniz (1989, cf. 1875–1890, IV: 608–609 for the untranslated text).

[14] Leibniz, *New System of the Nature and Communication of Substances* cited from Leibniz (1989, 144, cf. 1875–1890, IV: 485 for the untranslated text).

[15] Leibniz, *Letter to De Volder 1704/05* cited from Leibniz (1989, 181, cf. 1875–1890, II: 275 for the untranslated text).

1.2 The Living Individual with Perception and Appetite as the "Atom": The Fundamental Element of Nature

Leibniz restored their governing and perceiving souls to animals and even to plants. All living things, according to a text of 1690, possess not only a body, but also a soul, a "single indivisible substance [...] by which the animal or plant [...] is controlled."[16] This claim was upheld in the *New Essays* of 1704. "The great analogy between plants and animals inclines me to believe that there is some perception and appetite even in plants."[17] Moreover, no part of nature was purely corporeal. According to a well-known passage, even seemingly inert substances like earth and stone were densely inhabited by living beings.

> Each portion of matter "can be conceived as a garden full of plants, and as a pond full of fish [...] And although the earth and air lying between the garden plants, or the water lying between the fish of the pond, are neither plant nor fish, they contain yet more of them, though of a subtleness imperceptible to us."[18]

These ensouled creatures are all active: "Not only do immaterial substances always exist, but also that their lives, progress, changes, are always directed toward a definite end."[19] Each substance, each living individual, is different from every other and the creation is the set of all compossible possible substances.[20]

---

[16] Leibniz, *Comments on Michelangelo Fardella* cited from Leibniz (1989, 104). By "controlled" Leibniz did not mean to imply soul-body interaction.

[17] Leibniz, *New Essays* cited from Leibniz (1981, 139, cf. 1875–1890, V: 126 for the untranslated text).

[18] Leibniz, *Monadology*, §§ 67–68 cited from Leibniz (1989, 222, cf. 1875–1890, VI: 618 for the untranslated text).

[19] Leibniz, *Letter to Queen Sophie on What is Independent of Sense and Matter* cited from Leibniz (1989, 192, cf. 1875–1890, VI: 507–508 for the untranslated text).

[20] Leibniz, *My Principle is: Whatever can Exist and is Compatible with Others, Exists* cited from Leibniz (1992, 105).

1.3 The Tendency to Perfection

During his lifetime, Leibniz advanced numerous proposals for the improvement of law and litigation, for industry, public health, and the organization of science.[21] His metaphysics encompassed both the claim that the world God created is perfect in that God has selected the best and largest set of individuals possible, and the claim that there is historical progress towards better and better states culturally, morally, and politically. These claims were not incompatible, as he explained: "[T]hough the state of the world could never be absolutely perfect at any particular instant whatever [...] nevertheless the whole actual sequence would always be the most perfect of all possible sequences."[22]

In the "Radical Origination of Things," a work not published until the 1840s, he stated:

> [W]e must also recognize that the entire universe is involved in a perpetual and most free progress, so that it is always advancing toward greater culture. Thus a great part of our earth has now received cultivation and will receive it more and more. And though it is true that some sections occasionally revert into wilderness or are destroyed and sink back again, this must be understood in the same sense in which I have just explained the nature of afflictions, namely, that this very destruction and decline lead to a better result, so that we somehow gain through our very loss.[23]

At the same time, "There always remain in the abyss of things parts which are still asleep. These are to be aroused and developed into something greater and better, and in a word to a better culture. And hence progress never comes to an end."[24]

The doctrine of historical progress as unfolding on its own as humanity gained in age and experience was reinforced by Leibniz's theory of

---

[21] As is emphasized by Antognazza (2009).

[22] Leibniz, *Letter to Louis Bourguet, 1715* cited from Leibniz (1969, 664, cf. 1875–1890, III: 583 for the untranslated text).

[23] Leibniz, *On the Radical Origination of Things* cited from Leibniz (1969, 490–491, cf. 1875–1890, VII: 308 for the untranslated text).

[24] Leibniz, *On the Radical Origination of Things* cited from Leibniz (1969, 490–491, cf. Leibniz 1875–1890, VII: 308 for the untranslated text).

generation. According to the theory of preformation which he sought to popularize, the future organism resided invisibly, not as a miniature model of its later self, but enfolded in the seed, or egg or animalcule and requiring only to be released from its constraints to develop spontaneously.[25]

1.4 The Theory of Indistinct Knowledge, the Unconscious, and Its Motivating Force and Creative Power: Innateness

Temperamentally, Leibniz was not inclined to the supposed virtues of tranquillity and patience. Happiness does not consist in "complete joy, in which nothing is left to desire, and which would dull our mind" but in "a perpetual progress to new pleasures and new perfections."[26]

In discussing the doctrine of the "unease" that Locke thought motivated all our actions, including both moral and amoral behaviour, Leibniz agreed that volition could not exist without desire and aversion. The mind seeks equilibrium, but is constantly thrown out of it, mechanically perturbed by "some new effect of objects, some small change of the sense organs, and in the viscera and bodily cavities."[27] Our impulses "are like so many little springs trying to unwind and so driving our machine along."[28] Not only in fear, anger, envy, and shame are we motivated by disquiet, but even in hope, calmness, generosity, and pride.[29]

Where Locke maintained that unease was always felt as disagreeable, Leibniz argued that motivation may arise from "minute insensible perceptions which could be called sufferings that we cannot become aware

[25] Leibniz, *Theodicy*, § 90 cited from Leibniz (1985, cf. 1875–1890, VI: 152 for the untranslated text); Leibniz, *Principles of Nature and Grace* § 6 cited from Leibniz (1989, 209, cf. 1875–1890, VI: 601 for the untranslated text).

[26] Leibniz, *Principles of Nature and Grace* cited from Leibniz (1989, 213, cf. 1875–1890, VI 332–336 for the untranslated text).

[27] Leibniz, *New Essays* cited from Leibniz (1981, 166, cf. 1875–1890, V: 153 for the untranslated text).

[28] Leibniz, *New Essays* cited from Leibniz (1981, 166, cf. 1875–1890, V: 153 for the untranslated text).

[29] Leibniz, *New Essays* cited from Leibniz (1981, 192–193, cf. 1875–1890, V: 177–178 for the untranslated text).

of."[30] Healthy appetite, which is not unpleasant, is to be distinguished from "a tormenting obsession with the idea of whatever it is we are without."[31] Unconscious perceptions could even discriminate between seemingly identical objects and rouse a preference for one over the other. Knowledge, of which we are currently unaware, resides within us, and can therefore guide us in our actions; a goal need not be clearly or even consciously represented in order to be the object of striving.

In the *New Essays*, the mind is described "like a magic lantern" that produces unbidden (erotic) images, and in the *Theodicy*, the creative unconscious is described as analogous to the developmental and formative forces of nature.

> What necessity is there for one always to be aware how that which is done is done? Are salts, metals, plants, animals and a thousand other animate or inanimate bodies aware how that which they do is done, and need they be aware? Must a drop of oil or of fat understand geometry in order to become round on the surface of water? Sewing stitches is another matter: one acts for an end, one must be aware of the means. But we do not form our ideas because we will to do so, they form themselves within us, they form themselves through us, not in consequence of our will, but in accordance with our nature and that of things.
>
> The foetus forms itself in the animal, and a thousand other wonders of nature are produced by a certain instinct that God has placed there, that is by virtue of divine preformation which has made these admirable automata, adapted to produce mechanically such beautiful effects. It is easy to believe that the soul is a spiritual automaton still more admirable, and that it is through divine preformation that it produces these beautiful ideas, wherein our will has no part and to which our art cannot attain.[32]

"In sleep there often come to me," he writes in an early fragment, "pictures of great buildings which I have never seen, while in waking it would be difficult for me and I could succeed only with enormous difficulty in

[30] Leibniz, *New Essays* cited from Leibniz (1981, 188, cf. 1875–1890, V: 174 for the untranslated text).

[31] Leibniz, *New Essays* cited from Leibniz (1981, 189, cf. 1875–1890, V: 175 for the untranslated text).

[32] Leibniz, *Theodicy* cited from Leibniz (1985, cf. 1875–1890, VI: 403 for the untranslated text).

framing the idea of the simplest house. [...] [I]f only I could have held onto all the wonderful speeches, books, letters, and moving poems, which I have never read but have encountered in my dreams, where they seem to come into being only at the moment in which they present themselves to the eye."[33]

The theme of artistry driven by unconscious processes recurs repeatedly. In the *Principles of Nature and Grace*, Leibniz compares the mind's creative power to God's. "The mind not only has a perception of God's works, but it is even capable of producing something that resembles them, although on a small scale. For to say nothing of the wonders of dreams in which we effortlessly (though also involuntarily) invent things which we would have to ponder long to come upon when awake, our soul is also like an architect in its voluntary actions; and in discovering the sciences [...] it imitates in its realm and in the small world in which it is allowed to work what God does in the large world."[34]

1.5 Positive Reception of Leibnizianism

The tributes to Leibniz in the last quarter of the eighteenth century refer repeatedly to natural forces, the powers of the soul, and to nature's productivity and creativity. Herder enthusiastically cited Leibniz on the *Wirkende Kraefte der Natur*. "All active forces of nature are living ones; and there must be something within [the organism] that corresponds to their effects seen from without, as Leibniz divined and taught us. That we have no names for the inner organisation of the plant or the forces working in it, is only a deficiency in our language."[35] Further, "[N]o force of nature is without its organ [...]. Wherever we see a force at work, it is in an organ and acting harmoniously."[36] The "inner organic forces" of nature build the snail's shell, the bee's cells, and the spider's web.[37] Plants possess

[33] Leibniz, *De somnio et vigilia* cited from Leibniz (1923–, II: 277).

[34] Leibniz, *Principles of Nature and Grace*, § 14 cited from Leibniz (1989, 211–212, cf. 1875–1890, VI: 604–605 for the untranslated text).

[35] Herder (1989a, 101).

[36] Herder (1989b, 171).

[37] Herder (1989b, 171).

a generative drive which produces the "*Kunstwerke*" of seeds and flowers in addition to their nutritional drive.[38] Herder finds continuity in nature from minerals to plants to animals, hearkening back to the *Monadology*'s imagery of a full and maximally varied world and the controversial letter on the *Scala Naturae*.[39] For Herder:

> However great the variety of terrains and climates, of the stones, and plants of our earth, how much greater is the differentiation of its living inhabitants. They are not limited to the earth; air, the water, even the inner parts of plants and animals throng with living things. [...] The world's lively surface upon which everything, as far and wide as the sun reaches, enjoys, acts, and lives.[40]

For Schelling, the philosophy of Nature is founded on a "dynamic atomism." It assumes "an original multiplicity of individual principles," not as "actual material parts" but rather as "original and simple activities."[41]

The progress debates of the late eighteenth century set the pessimists Bayle, Mendelssohn, and Robinet, who saw only an oscillation between better and worse states, or a Manichaean state of constant opposition, against the optimists Leibniz, Lessing, and Kant, even if for the last named, hopefulness was a moral duty, a form of faith encouraged by certain signs.[42] Although Herder expressed himself sarcastically with regard to conquest, Europeanization, and the "eternal striving" of the human race, and although he denied that the growth of culture had brought about an increase in the virtue and happiness of mankind, the analogy between the growth and maturation of organisms and the growth and maturation of the human species could be seen, thanks to Leibniz, as natural and irreversible processes.[43]

---

[38] Herder (1989a, 101).

[39] This influential letter, allegedly by Leibniz, proposed a Great Chain of Being extending from minerals through plants to humans and angels. It was long believed to be a forgery but see the defence of its authenticity in Goldenbaum (2016).

[40] Herder (1989c, 69). Cf. Leibniz, *Monadology* §§ 66–69.

[41] Schelling (2004, 21n).

[42] Kant (1908c, AA VIII, 25–28).

[43] Herder (2004, 30–31).

2 The Other Aspect of the Leibnizian Legacy

The dominant impression of the Leibniz-Wolff philosophy in Germany before the publication and uptake of the *New Essays* was metaphysically static rather than biologically and historically dynamic. The components of Leibniz's philosophy that were not favourable to *Naturphilosophie* or the development and differentiation of notions of drive included the following:

1. "Metaphysics and Theology"
2. "Mechanism and Pre-established Harmony"
3. "Neglect of Animal Psychology and Agency"

2.1 Metaphysics and Theology

While Leibniz's physics and later his physiology of sensation in the *New Essays* discussed interactions between bodies, his metaphysics and his theory of life describe the monads as "windowless," as unaffected by one another. It is to each substance as though it were alone in the world with God and even the famous image of the fishpond reinforces the sense of independence. Fish swim in the pond and the water between them is filled with smaller organisms, and the water between them with even smaller organisms and so on. The scene is one of peaceful plenitude with room for all.

Despite Leibniz's rejection of tranquillity as an ethical ideal, his claim that, whatever is happening, everything is as good as it could be at the moment was not subject to nuanced interpretation and was much ridiculed. The leading philosophers of the eighteenth century in France and Britain, influential in Germany, were inclined to religious scepticism. Their images of nature and human society portray the world as theatre of complex interactions and forces working in opposition. The realization that God is not the lawgiver and not in charge of human destiny implied that human beings must work out their morals and politics for themselves. They had to reckon with human irrationality and malice not as the theological problems of why God permits sin, but as challenges for the parent, the educationalist, and the philosopher. This required a different perspective on the role of forces and drives in human beings.

2.2 Mechanism and Pre-established Harmony

Leibniz expressed a Cartesian hostility to nonmechanical agents and principles and forces active in the phenomenal world. The Leibnizian animal is a machine that is "a machine in all its smallest parts to infinity."[44] He attributed to mechanism "everything which takes place in the bodies of plants and animals except their initial formation."[45] Too many clever people, he complained, reject the correct view that everything that happens does so according to number, weight measure, size, shape, and motion, that is, mechanically. People imagine in animals "certain inner guardians, which are stimulated, calmed, excited or dulled, as circumstances arise, some of which are particular supervisors of certain limbs or internal organs."[46] In his polemic with Stahl, he attacked the biomedical soul that Stahl thought prevented decay during the life of the organism.[47] Even gravity, elasticity, and magnetism must be reducible to mechanisms. Attacking "barbaric physics," Leibniz assigned to the pathetic category of "renovators of scholastic qualities" everyone whose doctrines he disapproved of, throwing into the same bucket Newtonian attraction and repulsion, plastic intelligences directing the formation of the foetus, Paracelsian elemental chemistry, the soul of the world, and operative ideas.[48]

---

[44] Leibniz, *Monadology* § 64 cited from Leibniz (1989, cf. 1875–1890, VI: 618 for the untranslated text).

[45] Leibniz, *New Essays* cited from Leibniz (1981, 139, cf. 1875–1890, V: 126 for the untranslated text).

[46] Leibniz, *Against Barbaric Physics* cited from Leibniz (1989, 314, cf. 1875–1890, VII: 338 for the untranslated text).

[47] Cf. Leibniz (2016) for a translation of the Leibniz-Stahl Controversy.

[48] Leibniz, *Against Barbaric Physics* cited from Leibniz (1989, 319–320, cf. 1875–1890, VII: 337–344 for the untranslated text).

A consequence of his mechanism and his theory of individuals, the issue of determinism had bedevilled Leibniz since he first laid out to Arnauld his theory of the "complete concept," his idea that a person "contains" all his or her future states as the truth-makers of later statements about him or her. As he stated unapologetically in the *Theodicy*:

> All is therefore certain and determined beforehand in man, as everywhere else, and the human soul is a kind of spiritual automaton, although contingent actions in general and free action in particular are not on that account necessary with an absolute necessity, which would be truly incompatible with contingency.[49]

According to the *New System of the Nature and Communication of Substances* of 1695, the sequence of mental events of thoughts, feelings, and desires ran parallel to changes in the state of brain and body without causally influencing them. The soul could represent a future state as a goal, but the feeling of striving towards it—a perception—could not be causally efficacious in any movements of the body, and was merely correlated with the changes in the body that were preparatory to or constitutive of its movements.

Leibniz's interest in robotics was evinced in a notable passage in which, reacting to Bayle's supposition that the idea of a self-steering ship is absurd, he argues that this could not be done by means of an occult quality even conferred by God but "if [Bayle] means a faculty of the ship which is explicable by mechanical rules" that is another matter.

> There is no doubt that a man could make a machine which was capable of walking around a town for a time, and of turning precisely at the corners of certain streets. And an incomparably more perfect, although still limited mind could [make a machine to] foresee and avoid an incomparably greater number of obstacles. And this being so, if this world were, as some think it is only a combination of a finite number of atoms which interact in accord with mechanical laws, it is certain that a finite mind could be sufficiently

[49] Leibniz, *Theodicy* § 52 cited from Leibniz (1985, cf. 1875–1890, VI: 131 for the untranslated text).

exalted as to understand and predict with certainty everything that will happen in a given period.[50]

The expulsion of the Leibnizian professor, Christian Wolff, from his chair in Halle in 1723 turned on his determinism and optimism. Wolff's pietist opponents, with their concepts of sinfulness and responsibility, could not brook his metaphysics and theology.[51] The understanding of Leibniz as, along with the notorious Julien Offray de La Mettrie, a proponent of "machine man" is reflected as late as 1784 in Kant's claim that "if a human being's actions insofar as they belong to his determinations in time were not merely determinations of him as appearance but as a thing in itself, freedom could not be saved. A human being would be a marionette or an automaton, like Vaucanson's, built and wound up by the supreme artist."[52] In the *Critique of Practical Reason*, he compared the "spontaneous" action of a Leibnizian monad with the "freedom of a turnspit, which, when once it is wound up, also accomplishes its movements of itself."[53]

2.3 Neglect of Animal Psychology and Agency

Leibniz was more interested in the phenomena of perception than the phenomena of deliberate agency in animals and humans. Perceptual phenomena discussed at length included perspectivalism; the "mirroring" of the monads; the generation of new qualities by indistinct perception; subliminal perception; and the revelations of the microworld accessible only with optical instruments. Although he ascribed appetition and so goals in dim form to all living entities, he had no interest in animal behaviour as such. Where Descartes' theory of the living animal was interactive—he discusses predator-prey behaviour and seems familiar with the behaviour of dogs—Leibniz's references to nonhuman animals

[50] Leibniz, *Reply to the Comments in the Second Edition of M. Bayle's Critical Dictionary in the Article 'Rorarius' concerning the System of Pre-established Harmony* cited from Leibniz (1997, 108–109, cf. 1875–1890, IV: 555–556 for the untranslated text).

[51] See Langhansen (1724).

[52] Kant (1908a, AA V, 101), translated as Kant (1996, 221).

[53] Kant (1908a, AA V, 97), translated as Kant (1996, 218).

are entirely generic, centred on their souls, their mechanical nature, their immortality, their irrationality, and their generative abilities.

3 Leibniz-Reimarus *Naturphilosophie*

I return to Wilm's claim for Leibniz as an inspiration for *Naturphilosophie*. To repeat, Wilms credited Leibniz with initiating a tradition of "voluntarism" which reached into the nineteenth century mediated via Reimarus's views on animal *Triebe*.

It is hard to credit Leibniz with initiating this tradition. Dynamic concepts have always held an important place in natural philosophy, frequently, though not always in reaction to versions of the mechanical philosophy. In this connection one might mention Aristotle's *physis* or energy of nature, that behaves as though it possessed intelligence, though it does not; the Stoic doctrine that every animal strives for self-preservation, Strato's plastic and seminal powers, and the *conatus* of the mediaevals. The *appetitus* of late scholastic philosophy applies both to psychology and to physics.[54] Formative and teleological agents abound in Renaissance philosophy, including the indwelling active *Kraefte* of Parcelsus, Van Helmont's *archaeus*, and Kepler's *vis formatrix*. In the seventeenth century, there are More's "hylarchic principles" and Cudworth's "plastic nature."[55] As noted, Leibniz explicitly scorned most of these purported entities. He traced (or preferred and purported to trace) his own employment of the *conatus* concept to his teacher Erhard Weigel, rather than to Hobbes, Kepler, or Spinoza, in whom it is prominent.[56]

---

[54] See Garau (2014). Garau calls attention to the psychologization of nature or at least the explicit analogies between physical tendencies and animal and human appetites in the mediaevals (cf. Garau 2014, 481–489).

[55] For example, Ralph Cudworth, whose *The True Intellectual System of the Universe* (1678) appeared in Latin in Germany in 1733, translated by J. L. Mosheim. "Moreover, that something may act artificially and for ends, without comprehending the reason of what it doth, may be further evinced from those natural instincts that are in animals, which without knowledge direct them to act regularly, in order both to their own good, and the good of the universe" (Cudworth as quoted in Cheung and Cheungs 2006, 144n).

[56] Loemker (1946).

The specifically eighteenth-century view of Nature as an ensemble of independent but simultaneously acting, empirically demonstrable forces derives less from the history of philosophy than from the burgeoning experimental studies of electricity, magnetism, chemical affinity, and mesmerism. It is better credited to Leibniz's opponent Isaac Newton. Already in 1675, Newton had described nature as "a perpetual circulatory worker,"[57] speculating in an exciting way in his later *Opticks* on a range of active principles. The Comte de Buffon echoed this language in his 1764 essay, "Nature: A First View." "Nature is herself a work perpetually alive, an active and unceasing operator, who knows how to make use of every material [...]. [...] Those springs which she makes use of are active forces which time and space can only limit but never destroy, forces which unite, balance and oppose."[58] The materialist and atheist Holbach described matter as "subsumed under the empire of diverse attractive and active forces."[59]

Where did Reimarus, a dedicated opponent of revealed religion, but a leading deist, fit into the picture? Reimarus's main target was the neo-Epicureans who proposed—in however veiled fashion—that the world and its animals had arisen by chance and the perishing of the unfit, a speculation dating back to Empedocles and Lucretius.[60] In his *Vornehmsten Wahrheiten der Natuerlichen Religion* (1756, 7th ed., 1798), Reimarus presented a general theory of nature that he claimed implied indisputably the existence and creative activity of "a real, self-existent, and eternal Being."[61] Nature is the totality of "the essential power of every real substance." This applies to the material world including planets, stars, plants, and minerals. The whole world consists of "powers that act against each other" that are brought into congruity by a "wise connection."[62]

In Chapter III, Reimarus drew an unLeibnizian line between the "material world," devoid of life, sensation, and intelligence, and the

---

[57] Isaac Newton, Letter to Oldenburg (7 Dec 1675), quoted McGuire (1968).

[58] Leclerc and de Buffon (1797, 326).

[59] Thiery and von Holbach (1770, II: 322).

[60] The English translator's subtitle was: 'Wherein the Objections of Lucretius, Buffon, Maupertuis, Rousseau, La Mettrie, and other ancient and Modern Followers of Epicurus Are Considered.'

[61] Reimarus (1766, 115).

[62] Reimarus (1766, 137).

world's populations of "men and animals." The former consisted of the celestial bodies, geographical features, fire, water, atoms, rocks, and plants. All the changes in the material world, which Reimarus compared—here however in Leibnizian fashion—to clockwork, "are produced according to fixed rules." The material universe is a "vast machine [whose] suns and planets" are its "parts and wheels."[63]

Like Leibniz, Reimarus explicitly denied that there are plastic natures, hylarchic principles, or formative ideas.[64] Inanimate bodies have powers but no consciousness of them. The material world was produced by a divinity only for the sake of the living beings who observe and enjoy it and who benefit from its features.[65] The whole corporeal world is unconsciously "continually at work for our benefit."[66]

Reimarus denied "the pretended perfection which some attribute to nature from the universal life bestowed on its minutest parts."[67] Some philosophers, he remarked, "chose to ascribe life to every particular substance, animating every plant and even the smallest particle of matter. This hypothesis has at first sight an air of grandeur and dignity; since life is thereby multiplied and all nature animated." But it is false: plants and atoms lack life, sensation, thought and volition.[68] A world in which the planets move in their regular courses, and trees, grain, and flowers flourish is still "solitary, silent, dead."[69]

Reimarus's dualism and subjectivism excluded a Leibnizian attribution of metaphysical perfection to the entire world and foreshadowed Kant's "Copernican twist." The perfection of any lifeless thing—natural or artificial—depends on its being perceived by a sensitive being. The beauty and scent of flowers, for example, is not an "intrinsic perfection" of theirs since they cannot perceive or enjoy it. They, along with the rest of the dead world have no power to achieve their "natural perfection," which

[63] Reimarus (1766, 95).

[64] Reimarus (1766, 92).

[65] Reimarus (1766, 120).

[66] Reimarus (1766, 143).

[67] Reimarus (1766, 94).

[68] Reimarus (1766, 93).

[69] Reimarus (1766, 105). Confusingly, Reimarus appears to add elephant, mouse, and whale to the "horrid and dreary" category of dead and inert things.

Reimarus describes as the attainment of pleasure, happiness, and life in accord with the nature of the species.[70] All animate bodies, however, have "a sense of their existence" their aims and powers. The animal must be organized in its structure and functioning for the pursuit and likely attainment of its particular form of wellbeing.

> [A]ll animate Beings have an internal end attainable by their natural powers, which they are continually endeavouring to arrive at; and everything that is in them corresponds and harmonizes with this end, and is determined by it [...] [B]y virtue of their own natures, they endeavour to make themselves perfect, and enjoy their perfection in pleasing sensations.[71]

> [T]he final cause of the natural powers of animate beings is in themselves; and accordingly, this inward design or tendency of different animals to some end determines what powers, inclinations, capacities, and actions, what members and organs of the body, every species of living creatures must essentially have; and, how each of them must be constructed and disposed to correspond with its final cause, that is, with its peculiar manner of life, the nature of its affections, and with its pleasure or happiness; in a word, that it may have its Natural Perfection.[72]

Reimarus's views on animal instincts in the various genera or "species" were developed in his *Betrachtungen über die Triebe der Thiere* (1760, 4th ed., 1798). This text will be dealt with by other contributors to this volume, but I summarize briefly: The capability and industry of nonhuman animals—especially wild animals whose nature had not been deformed into passive dependence and stupidity by domestication—were studied in the eighteenth century with a new breadth and intensity. From birth, it was observed, birds and mammals recognized their conspecifics, were attracted to certain foods and avoided others, and performed the behaviour necessary for finding and attracting mates and for raising their young. They possessed the architectural knowledge needed for building

---

[70] Reimarus (1766, 96).

[71] Reimarus (1766, 97–98).

[72] Reimarus (1766, 98).

hives, nests, and dams. Knowledge, ability, and drive must have been impressed into them at the Creation.

Reimarus criticized Descartes, for his beast-machine doctrine, and, amongst the "moderns," Buffon for attributing actions and experiences to "certain vibrations of their organs and the inward texture of their brains."[73] Though he admitted that Buffon ascribed "dreams, experience, imitation, pleasure and pain, inclinations and aversion" to nonhuman animals, he presented a fork: if animals are purely mechanical, only God could have constructed such marvellous machines; and if they are ensouled, only God could have endowed them with their souls.[74]

Despite their shared providential framework and their shared interest in vitality, Reimarus was as critical in certain respects of Leibniz as he was of Descartes.[75] As he said of the latter: "Leibniz and Wolff, the great men who wrote so beautifully about forces, did not address the comparison of the various forces in the whole of nature or the different levels of their determination."[76] Wolff, he observed correctly, treated only the human soul in his *Psychologia Empirica*, and the same was true of Leibniz. The system of pre-established harmony between soul and body was incompatible with Reimarus's belief that an internal, incorporeal principle in animals governed their behaviour. Leibniz's system was only a slight improvement on the Cartesian beast machine, he thought. "It restored to the poor animals their souls and lives, their feelings and imagination." But the Leibnizian soul could not put the body in motion:

> When one examines this clever hypothesis more closely, the animal body remains essentially a plain Cartesian machine that is not enlivened by a soul and does not receive any influence from one. So that even with the soul, one does not have a living animal; but only something that moves around according to what the soul perceives and that would do the same as it does even without a soul.[77]

---

[73] Reimarus (1773, 238–240). On Reimarus and Kant, see Engfer (1980).

[74] Reimarus (1766, 236–240).

[75] Though see Leduc (2020), for an account of Reimarus's reworkings of the Leibnizian themes of harmony and teleology.

[76] Reimarus (1773) "Anhang von der verschiedenen Determinationen der Naturkraefte," 411.

[77] Reimarus (1760, 221–2).

4 Conclusion

Leibniz was read and appreciated by romantic philosophers, among them Kant's critic, J.A. Eberhard, Herder, and Schelling.[78] "His mind," said Schelling, "despised the fetters of the schools; small wonder that he has survived among us only in a few kindred spirits and among the rest has long become a stranger. He belonged to the few who treat science as a free activity. He had in himself the universal spirit of the world, which reveals itself in the most manifold forms, and, where it enters, life expands."[79] Judging from the absence of broad enthusiasm in Germany for Leibniz's metaphysics, and from the availability of so many other forms of philosophical dynamism, the claim for a Leibnizian stamp on theories of nature in later eighteenth- and nineteenth-century thought is somewhat contentious.

Wilm and later historians of philosophy were right to identify Leibniz as the preeminent seventeenth-century critic of Descartes' theory of substance as powerless and inert. No early modern metaphysician undertook as systematically and repetitively as Leibniz did to present physical, developmental, and psychological forces as analogous, while remaining committed to a strict version of the mathematical-mechanical philosophy. After the publication of the *New Essays*, Leibniz's views on human volition and his theory of unconscious perception and unconscious appetite were available for appreciation.[80] And although he had not a great deal to say about artistic creativity and invention, what he had to say on these topics was intriguing and profound. The analogy between introspected psychological forces and measurable and observable physical and developmental forces was not new in Leibniz, but it lay well apart from the Newtonian experimental tradition. And it was stimulating to those later philosophers who invested the universe with unity and meaning. Schelling and Fourier considered sexual attraction as having cosmological

---

[78] Wilson (1994).

[79] Schelling (1989, 16).

[80] The first German translation appeared only in 1778, in a translation of Eric Raspe's 1765 collection of Leibniz's writings. On the reception of the work, see Tonelli 17,974 and on its uptake Whyte (1962).

significance, and the chemistry-derived theme of irresistible erotic attraction is the subject of Goethe's 1809 novella, *Elective Affinities*.

Yet for the historian of the life and physical sciences, Leibniz is better classed amongst the opponents of the eighteenth-century and early nineteenth-century image of nature as field of harmonious and conflicting empirically real, law-governed forces, including vital and formative ones. His preformation theory was defeated in favour of an epigenetic account of generation, and this required reference to the building forces of nature that he had scorned. The principle of life, Herder decided, is a kind of ethereal or electrical current that circulates "in the vessels of plants, in the arteries, veins and muscles of animals, and finally in the nervous apparatus." Schelling referred to a "Life-Force" which, in the fashion of Leibniz's opponent Stahl, runs the organism and resists the decay that would otherwise overtake it.[81]

Leibniz's aim was to refute materialism and Spinozism while retaining the mechanical philosophy as uniquely explanatory. Although he depicted the individual as a locus of future-directed energy, his framework remained theological. He wanted to disarm religious scepticism while excluding God's direct intervention into created nature. But he would not go so far as to make nature self-sufficient: to create an "idol of nature." These aims were largely shared by Reimarus. By contrast, the late eighteenth- and early nineteenth-century romantic philosophers and proponents of *Naturphilosophie* saw no need to mix divinity with natural philosophy. They saw metaphysical idealism and the supernatural as a poor response to the perceived threat of a soulless materialism and the re-enchantment of nature as the better way forward. They benefitted in this reconstruction from nearly a century of investigation, after Leibniz's time, into the physical forces of nature and the drives motivating animals and humans.

---

[81] Schelling (1989, 37).

References

Antognazza, Maria Rosa. 2009. *Leibniz: A Biography*. Cambridge: Cambridge University Press.

Beiser, Frederick. 2000. The Enlightenment and Idealism. In *The Cambridge Companion to German Idealism*, ed. Karl Ameriks, 18–36. Cambridge: Cambridge University Press.

Cheung, Tobias, and Thomas Cheungs. 2006. Hermann Samuel Reimarus' Theorie der 'Lebensarten' und 'Triebe'. *Sudhoffs Archiv* 90 (2): 143–160.

Engfer, Hans-Jürgen. 1980. Die Urteilstheorie von H. S. Reimarus und die Stellung seiner 'Vernunftlehre' zwischen Wolff und Kant. In *Logik im Zeitalter der Aufklärung. Studien zur "Vernunftlehre" von Hermann Samuel Reimarus*, ed. Walter Wolf and Ludwig Borinski, 33–58. Göttingen: Vandenhoeck & Ruprecht.

Garau, Rodolfo. 2014. Late-scholastic and Cartesian conatus. *Intellectual History Review* 24 (4): 479–494.

Goldenbaum, Ursula. 2016. *Ein Plädoyer für die Echtheit des angeblich gefälschten Leibnizbriefes*. Hannover: Wehrhahn.

Herder, Johann Gottfried. 1989a. Von den Trieben der Tiere. In *Ideen zur Geschichte der Menschheit, Johann Gottfried Herder*, ed. Martin Bollacher, 100–105. Frankfurt am Main: Deutscher Klassiker Verlag.

———. 1989b. Keine Kraft der Natur ist ohne Organ. In *Ideen zur Geschichte der Menschheit, Johann Gottfried Herder*, ed. Martin Bollacher, 171–175. Frankfurt am Main: Deutscher Klassiker Verlag.

———. 1989c. Das Reich der Tiere. In *Ideen zur Geschichte der Menschheit, Johann Gottfried Herder*, ed. Martin Bollacher, 67–76. Frankfurt am Main: Deutscher Klassiker Verlag.

———. 2004. *Another Philosophy of History for the Education of Mankind*. Trans. and Ed. Ioannis D. Evrigenis and Daniel Pellerin. Indianapolis: Hackett.

Jaynes, Julian, and William R. Woodward. 1974a. In the Shadow of the Enlightenment: I. Reimarus Against the Epicureans. *Journal of the History of the Behavioral Sciences* 10 (1): 3–15.

———. 1974b. In the Shadow of the Enlightenment: II. Reimarus and His Theory of Drives. *Journal of the History of the Behavioral Sciences* 10 (2): 144–159.

Kant, Immanuel. 1908a. Kritik der praktischen Vernunft. In *Gesammelte Schriften*, ed. Akademie der Wissenschaften, vol. V, 1–163. Berlin: Reimer.

———. 1908b. Kritik der Urteilskraft. In *Gesammelte Schriften*, ed. Immanuel Kant and Akademie der Wissenschaften, vol. V, 165–485. Berlin: Reimer.
———. 1908c. Ideen zu einer allgemeinen Geschichte in weltbürgerlicher Absicht. In *Gesammelte Schriften*, ed. Akademie der Wissenschaften, vol. VIII, 15–32. Berlin: Reimer.
———. 1996. Critique of practical reason. In *Practical Philosophy*. Trans. and Ed. Mary J. Gregor, 133–272. Cambridge: Cambridge University Press.
Langhansen, Christoph. c. 1721. *Dubia Circa Monades Leibnitianas Qvatenus Ipsae Pro Elementis. Corporum Venditantur*. Regiomonti: Reusner.
———. 1724. *De Necessitate Omnium, Quae Existunt, Absoluta, in Theodicaea G. G. Leibnitii, Cui Wolfianum Metaphysicae Systema Superstructum Est, Asserta*. Thesis: Universitaet Königsberg.
Leclerc, George-Louis, and Comte de Buffon. 1797. A First View of Nature. In *Buffon's Natural History. Containing a Theory of the Earth, a general History of Man of the Brute Creation, and of Vegetables, Minerals &c. &c*, vol. X, 325–242. London: H. D. Symonds.
Leduc, Christian. 2020. Reimarus on Natural Religion, Final Causation and Mechanism. *Studia Leibnitiana* 50 (1): 105–120.
Leibniz, Gottfried Wilhelm. 1720. *Des Hn. Gottfried Wilh. von Leibnitz Lehr-Sätze über die Monadologie*. Trans. Heinrich Köhlern. Frankfurt, Leipzig, Jena: J. Meyer.
———. 1765. *Nouveaux essais sur l'entendement humain par l'auteur du système de l'harmonie préétablie*. Ed. Eric Raspe, Amsterdam and Leipzig: Schreuder.
———. 1875–1890. *Die Philosophischen Schriften von G. W. Leibniz*, 7 vols. Ed. C.I. Gerhardt. Berlin: Weidmannsche Buchhandlung. (Reprint: Hildesheim: Olms, 1978).
———. 1923–. *Sämtliche Schriften und Briefe*. Ed. Deutsche Akademie der Wissenschaften. Darmstadt and Berlin: Akademie-Verlag.
———. 1969. *Gottfried Wilhelm Leibniz: Philosophical Papers and Letters*. Trans. and Ed. Leroy Loemker. Dordrecht: D. Reidel.
———. 1981. *New Essays on Human Understanding*. Trans. and Ed. Peter Remnant, and Jonathan Bennet. Cambridge: Cambridge University Press.
———. 1985. *Theodicy: Essays on the Goodness of God, the Freedom of Man, and the Origins of Evil*. Trans. E.M. Huggard and Ed. August Farrer. La Salle, IL: Open Court.
———. 1989. *Philosophical Essays*. Trans. and Ed. Roger Ariew, and Daniel Garber. Indianapolis: Hackett.

———. 1992. *De Summa Rerum: Metaphysical Paper, 1675–1676*. Trans. and Ed. G.H.R. Parkinson. New Haven: Yale University Press.
———. 1997. *Leibniz's New System and Associated Contemporary Texts*. Trans. and Ed. Roger Woolhouse and Richard Franks. Oxford: Clarendon Press.
———. 2016. *The Leibniz-Stahl Controversy*. Trans. and Ed. François Duchesneau, and Justin E.H. Smith. New Haven: Yale University Press.
Loemker, Leroy. 1946. Leibniz's Doctrine of Ideas. *The Philosophical Review* 55 (3): 229–249.
McGuire. 1968. Force, active principles, and Newton's invisible realm. *Ambix* 15 (3): 154–208.
Müller, Jakob Friedrich. 1745. *Die Ungegründete Und Idealistische Monadologie, Oder Wahre Gestalt Der Leibniz Und Wolfischen Lehre Von Denen Einfachen Dingen*. Franckfurt am Mayn: Eichenberg.
Ortlieb, Cornelia. 2010. Philosophie als Literaturkritik. Jacobi, Herder und Reimarus über die Triebe der Tiere. In *Friedrich Heinrich Jacobi und die Philosophie als Schreibart*, ed. Cornelia Ortlieb, 29–50. München: Wilhelm Fink.
Peterson, Keith R. 2004. Translator's Introduction. In *First Outline for a System of the Philosophy of Nature*, F.W.J. Schelling, Trans. and Ed. Keith R. Peterson, xi–xxxv. Albany, NY: State University of New York Press.
Reimarus, Herman Samuel. 1760. *Allgemeine Betrachtungen über die Triebe der Thiere, hauptsächlich über ihre Kunsttriebe: zum Erkenntniß des Zusammenhanges der Welt, des Schöpfers und unser selbst, vorgestellt von Hermann Samuel Reimarus*. Hamburg: Johann Carl Bohn.
———. 1766. *The Principal Truths of Natural Religion Defended and Illustrated in Nine Dissertations*. Trans. R. Wynne. London: B. Law.
———. 1773. *Allgemeine Betrachtigung ueber die Triebe der Thiere, hauptsachlich ueber ihre Kunst-Triebe*. 3rd ed. Hamburg: Johann Carl Bohn.
Schelling, Friedrich Wilhelm Joseph von. 1989. Introduction. In *Ideas for a Philosophy of Nature*, Friedrich Wilhelm Joseph von Schelling, Trans. Robert Stern. Cambridge: Cambridge University Press.
———. 2004. *First Outline for a System of the Philosophy of Nature*. Trans. and Ed. Keith R. Peterson. Albany, NY: State University of New York Press.
Thiery, Paul-Henri, Baron von Holbach. 1770/1994. *Systeme de la Nature ou Des Loix du Monde Physique et du Monde Moral*. Vol. 2. London. Reprint: Hildesheim: Olms.

Tonelli, Giorgio. 1974. Leibniz on Innate Ideas and the Early Reactions to the Publication of the Nouveaux essais (1765). *Journal of the History of Philosophy* 12 (4): 437–454.

Whyte, Lancelot. 1962. *The Unconscious before Freud*. New York: Doubleday.

Wilm, Emil Carl. 1925. *The Theories of Instinct: A Study in the History of Psychology*. New Haven: Yale University Press.

Wilson, Catherine. 1994. The reception of Leibniz in the 18th Century. In *The Cambridge Companion to Leibniz*, ed. Nicholas Jolley, 442–474. Cambridge: Cambridge University Press.

Zammito, John H. 2017. Herder between Reimarus and Tetens: The Problem of an Animal-Human Boundary. In *Herder*, edited by Anik Waldow and Nigel de Souza, 127–146. Oxford: Oxford University Press.

3

Between Reimarus and Kant: Blumenbach's Concept of *Trieb*

John H. Zammito

The concept *Trieb* played a central role in the developing life sciences of the late eighteenth century in Germany. The contrast between *Trieb* and *Kraft* (*nisus* and *vis*) highlighted the distinctive domain of life, as contrasted with more general physical-chemical forces in nature. Initially, the notion of force (*vis*) incorporated the emergent life sciences. Thus, at mid-century, Albrecht von Haller termed his two experimentally discriminated forms of animal physiology, irritability, and sensibility, *Lebenskräfte* (Haller 1755). This use of the word *Kraft* or its Latin cognate, *vis*, also characterized Caspar Friedrich Wolff's pioneering theory of generation: *vis essentialis* (Wolff 1966). The notion of *Lebenskraft* permeated late eighteenth-century discourse in life science in Germany (Engels 1994).

Only in that context can we appreciate the decisive terminological innovation of Johann Friedrich Blumenbach (1752–1840) in shifting from this notion of *Kraft* (*vis*—force) to the notion of *Trieb*

J. H. Zammito (✉)
Rice University, Houston, TX, USA
e-mail: zammito@rice.edu

M. Kisner, J. Noller (eds.), *The Concept of Drive in Classical German Philosophy*,
https://doi.org/10.1007/978-3-030-84160-7_3

(*nisus*—drive, but also, and seminally, *instinct*). For Blumenbach, the question of the emergence [*Entstehung*] of organized bodies was, despite all the efforts of modern science, "one of the most difficult topics of physiology." He elaborated: "Whence the *original basic material* [*erste Grundstoff*; italics in original] of each and every animal and plant arises [*hervorkomme*], and by what sorts of forces [*Kräfte*] this material is subsequently developed [*ausgebildet*], are each problems whose solution hitherto has been shrouded in darkness" (Blumenbach 1779, 17).[1]

With his greatest conceptual intervention, the *Bildungstrieb*, Blumenbach intentionally separated this phenomenon from the other *Bildungskräfte* that had been identified in the physical (and even the organic) world. Thus we must endeavor to reconstruct the source and the significance of Blumenbach's turn to *Trieb*, which proved important for Kant (and, later still, for Goethe and Schelling). My argument will be that Blumenbach drew upon the pioneering work of Hermann Reimarus; hence, it will articulate three stages: first, the conceptualization of *Trieb* in Reimarus; second, the introduction of that term into Blumenbach's own theorizing, in the notion *Bildungstrieb*; and, finally, briefly, how that figured in the reception of Blumenbach by Kant.

1 Reimarus and *Trieb* as "Instinct"

In *Allgemeine Betrachtungen über die Triebe der Thiere, hauptsächlich über ihre Kunsttriebe* (1760), Reimarus denied the central thesis of the Abbé de Condillac's *Traité des animaux* (1755) that animals learned through experience on analogy with human development (Condillac 2004). Instead, Reimarus insisted on the *innate* character of animal "instinct" or "drive" [*Trieb*] (Reimarus 1760, 319–321).[2] Reimarus built a powerful case for

[1] All translations from Reimarus and Blumenbach are my own.

[2] In his "Opening Remarks," Ernst Mayr makes the crucial point: "Condillac's book notably receives more pages of analysis than the work of any other author" (Mayr 1982, 14). Reimarus had a longer-standing interest in the technical question of animal instinct, and it was the convergence of this issue in life science with the larger questions of natural theology in the epoch that galvanized his work. Already as a student Reimarus had composed an essay, "*Instinctum Brutorum Existentis Dei Ejusdemque Sapientissimi Indicem*" (1725), reprinted and translated into German in the critical edition of Reimarus (1760, 757–780). These preoccupations in the German philosophical land-

the *innateness* of instinct, different *in kind* from human learning. "I demonstrate from the powers of animals and from the properties of the skillful drives [*Kunsttriebe*] that they do not consist in an effectiveness that the animals achieved for themselves through experience and reason, nor through a degree of reason, but that these are inherited capacities which arise out of the determinate natural powers of animals [*angeborne Fertigkeiten sind, welche aus den determinierten Naturkräften der Thiere entstehen*]" (Reimarus 1760, 66–67).

Already in his *Die vornehmsten Wahrheiten der natürlichen Religion* (1754), Reimarus offered a clear conceptualization of the essential features of the concept *Trieb*: it was the capacity to pursue a self-beneficial goal "without any individual reflection, experience and practice, without any training, example or model, from birth onward, with an artfulness ready from birth that was masterful in achieving its end" (Reimarus 1754, 379). He promised his readers that he would elaborate his ideas on instinct in a subsequent work and he fulfilled that promise in his *Triebe der Thiere* of 1760.[3] A major feature of the work was fascination with the question of the animal/human boundary in terms of the long-standing debate over the "animal soul." Based on extensive documentation articulating ten classes and fifty-seven subclasses of animal instinctual behavior, Reimarus achieved a massive synthesis of the data, giving his conception of *Trieb* a warrant that made it indisputable (Reimarus 1760, 65). He characterized "elective" [*willkürlichen*] drives as entailing "an inclination or aversion of the will towards present, though indistinct representations, that is toward the feeling of sensual pleasure or pain, whence the elective actions [*willkürliche Handlungen*] arise which are appropriate to the inclination or aversion" (Reimarus 1760,130). Among the elective drives, he distinguished between general self-preservation and more particular manifestations, which he labeled *Kunsttriebe* [skillful drives]. These were his primary concern and, indeed, the notion of *Kunsttrieb* proved the most theoretically fruitful concept he articulated. It was taken up by

scape were reflected in works by the key Wolffian philosopher at Halle, Georg Friedrich Meier (1742, 1747, 1749).

[3] As Jaynes and Woodward (1974a) point out, the whole of the *Triebe der Thiere* proved "an attempt to define the concept of drive." Disappointingly, the authors miss, glaringly, the connection with Johann Friedrich Blumenbach's crucial concept, *Bildungstrieb*.

Schiller and by Schelling, and, as we will note, it figured importantly in Blumenbach's notion of the *Bildungstrieb*. Reimarus began by characterizing *Trieb* in general as encompassing "all natural effort toward particular actions [*alles natürliche Bemühen zu gewissen Handlungen*]" (Reimarus 1760, 65, 78). He characterized as *Kunst* [artifice; skillful activity] all "regulated [i.e., rule-governed] capacities for certain actions" (Reimarus 1760, 147). Thus the decisive features of the *Kunsttrieb* were that it was natural and yet it was agential, that is, it entailed a measure of *purpose*. In short, Reimarus arrived at a notion of animal behavior as *immanent purposiveness*. It was *purposive* precisely in that it aimed "at the maintenance and well-being of the animal and its species [*auf die Erhaltung und Wohlfahrt des Thieres und seines Geschlechtes*]" (Reimarus 1760, 66). It was, however, altogether *innate*: This was his overriding theoretical insistence.

> How is it then conceivable that the animals, with such lowly physical and spiritual powers, in part without any external experience, without upbringing, guidance, examples, or verbal instruction, but above all without any reflection and actual thinking, without concepts, judgments, inferences and the discoveries that flow from these, without themselves knowing any purpose or the ability to recognize the relationship of means to that purpose, without experiments and long practice, nonetheless find themselves at all times capable of constructing completely and masterfully the most ubiquitously useful and clever artificial actions [*Kunsthandlungen*] for the many needs of every aspect of life and for the preservation of their species, and notably oftentimes from the very moment they come into the world? (Reimarus 1760, 364)

He answered: all *Triebe* were immediately "endowed by the Creator [*von dem Schöpfer eingepflanzt*]" (Reimarus 1760, 364).

Animals shared with humans certain features of sensibility—external sense organs, imagination, memory, an internal sense for pleasure and pain, and inclinations derived from these. Indeed, the "rule according to which the sensual representation of animals proceeds appears to be completely the same as the rule of our lower faculties of soul [*niederen Seelenkräfte*]" (Reimarus 1760, 381). He takes up in detail the work of the Halle philosopher G. F. Meier, *Versuche eines neuen Lehrgebäudes von den Seelen der Thiere* (1749), concentrating on the key school-philosophical distinction between the "higher" and "lower" faculties of mind, the key move in the rational psychology of the Wolffian school (Reimarus 1760, 336ff.). Still, the parallel had a different significance in the whole structure of awareness, in that animals were never capable of conceptual relation or inference. Reimarus worked through the levels of human reasoning—from concept-formation through judgment to inference—as developed in German school-philosophical teachings, and demonstrated that animals could achieve none of these. "If, then, all thought consists in concepts, judgments and inferences, then we cannot, in the literal meaning of the term, say that animals think" (Reimarus 1760, 121).

This was his key claim: "we can offer no grounds to ascribe reason or any grade of the same to animals, as contrasted with us humans, nor even to ascribe to their brains any natural images or innate figures which would be of use for this faculty [*einige diesem Vermögen behülfliche Naturbilder und angeborene Figuren in dem Gehirne beylegen*], or even to take into the account any extraordinary intervention by God" (Reimarus 1760, 363). Instead, they acted from conditioned reflex and habit. "It is from this, then, that we can grasp how the animals know things and discriminate among them, as well as how they are aware of what they are conscious of. Everything is merely indistinct and confused, and yet very lively" (Reimarus 1760, 107).[4] Humans operated discursively: It was from a general concept that they were enabled to discriminate instances of commonality (Reimarus 1760, 117). While humans classified conceptually, "the animal discrimination of species and genera has an entirely different basis and must be essentially different from our own" (Reimarus 1760, 110). Humans had the capacity for reflection, and reasoning made them truly human (Reimarus 1760, 158). That was instantiated in their unique capacity for language (Reimarus 1760, 123). He insisted that no animals, not even apes, could ever achieve language (Reimarus 1760, 126). But in animals, "sensibility in the vast majority of instances suffices to recognize and to distinguish individual things as well as types" (Reimarus 1760, 113). There were compensations in animals for lack of reflection: They

[4] These are the key categories of German school-philosophical faculty psychology (see Wundt 1939).

were by instinct far more effective in achieving the end of self-preservation. "Every animal expresses the creative drives of its species, from the very first trial, with a complete, regular competence, without previous instruction, exercise or mulling-over" (Reimarus 1760, 235). This efficacy was immediate, inveterate, and sufficient for its ends. "The same behaviors appear among these animals, from the start of their lives, a capacity which, without slow and awkward experimentation, without preliminary errancy and confusion, from the very first produces masterpieces" (Reimarus 1760, 170).

Reimarus insisted that a proportionality between efficacy and restriction was essential: "All the drives of animals are enclosed within the restrictions of their sensual awareness and desires" (Reimarus 1760, 222). "No species of animal has unnecessary or superfluous creative drives" (Reimarus 1760, 224). "No single animal has by nature alien, false or perverse creative drives" (Reimarus 1760, 225). Moreover, these drives were unchanging: "there arise just as few new arts among the animals as are lost or become weaker" (Reimarus 1760, 234).[5] Enhanced senses, exquisitely intricate and appropriate behaviors, and restricted but functional needs all worked to make animal drives efficacious far beyond anything that the considered and labored achievements of individual humans could attain. It took the generational accrual of implements and skills through the trial and error of rational reflection to bring humans to a superior estate in the natural order.

At the close of his chapter on the *Kunsttriebe*, Reimarus noted: "by the term skillful drives I designate the matter itself, which is before the eyes of everyone to observe, but not the cause or the manner of its possibility [...] One must first come to be acquainted with the matter itself according to its actual properties before one can ask how all that has really been observed in animals can be possible" (Reimarus 1760, 176). While the interaction between soul and body was ultimately a "mystery, and how it

[5] Reimarus is prepared to allow a degree of innovation to animal creative drives: "I will later show that the animals even in their creative drives do not behave in so completely uniform and mechanical a manner that one cannot concede to them the capacity to make different determinations here and there, according to circumstances" (Reimarus 1760, 110). Still, it was possible for animals to make discriminating use of their creative drives, under given circumstances (Reimarus 1760, 248). They could also err in the application of their instincts (Reimarus 1760, 256).

actually works will always be impossible for us to know," nevertheless, experience and observation made it clear that such interactions happened, both in animals and in humans (Reimarus 1760, 395). "We thereby must presume the actual connection between the soul and the body, merely according to experience, even if we cannot explain the nature of this mutual interaction" (Reimarus 1760, 406). This was the fundamental methodological maxim of empirical research science, Reimarus concluded:

> For the original essential forces are, in accordance with their regular determination, the first ground of all the actualities of nature: and all philosophers must acknowledge that it is impossible to determine *a priori* any further philosophical or mathematical proof concerning the fundamental forces of things and their determinate laws, but they must simply assume them in accordance with experience in order to carry on inquiry. (Reimarus 1760, 434)

Thus, Reimarus formulated *methodological-empirical* responses to crucial epistemological and ontological questions about the nature of forces and drives which bore upon human nature, not just the animals.

2 The Project of Johann Friedrich Blumenbach

In his influential *Handbuch der Naturgeschichte*, with numerous editions starting in 1779, Blumenbach explored all the conventional arguments whereby man was distinguished from the animals. His core distinction turned on behavior: Man had no instincts, but all the other animals were determined by them (Blumenbach 1779, 39–41). From the outset, in his terminology and argumentation Blumenbach clearly drew on Reimarus, though he would only cite him explicitly in later editions of the *Handbuch*.[6] He assigned humans to their own genus, separate from the apes, thus repudiating Linnaeus's inclusive class of primates (Linnaeus

[6] Blumenbach cited Reimarus explicitly starting in the third edition of the *Handbuch*.

1758; see Broberg 1994). Man was distinct precisely by virtue of his lack of instinct and organs of self-defense. His unique order was "*Inermis*—here taken specifically to betoken the absence of innate weapons, skillful instincts [*Kunsttriebe*—a technical term from Reimarus], cover [*Bedeckungen*—i.e., armor or camouflage]—in short all those [vulnerabilities] that man is rescued from by reason" (Blumenbach 1779, 57). In the *Handbuch* Blumenbach explicitly connected the absence of instinct in man with his ability to penetrate all habitats on earth and establish ubiquitous settlement.

Perhaps a key to dating Blumenbach's seminal publication on the *Bildungstrieb* was the death of his idol, Albrecht von Haller, early December 1777. Haller cast a very deep shadow over German life science; dissenting from his views was highly risky. Blumenbach had long since developed reservations about aspects of Haller's work, in particular concerning the theory of generation, but as long as Haller lived, Blumenbach never published them. Already while still a medical student, Blumenbach had been drawn to the most exciting experimental work in life science of the eighteenth century, namely Abraham Trembley's famous work on polyps.[7] In 1774 he made a presentation to the Göttingen Academy about his discovery of a new species of freshwater polyp in the environs of Göttingen.[8] This was the basis for experiments he then undertook in spring 1778 (Blumenbach 1780c). This culminated in late 1780, when Blumenbach published his most seminal scientific paper, "Über den *Bildungstrieb* (*Nisus formativus*) und seinen Einfluß auf die Generation und Reproduction," in a newly founded journal edited by his two colleagues, Georg Lichtenberg and Georg Forster, the *Göttingisches Magazin der Wissenschaften und Litteratur* (Blumenbach 1780b). The article title itself is important not only for its introduction of his famous term, with its Latin equivalent (*nisus*, not *vis*), but for its incorporation of propagation (German: *Generation*) and of regeneration (German: *Reproduction*) as essential aspects of the phenomenon. Blumenbach proposed that

[7] The importance of Trembley is well-captured in the title and substance of the work of Sylvia and Howard Lenhoff (1986).

[8] See Abraham Kästner's (1774) verbatim report of Blumenbach's presentation to the Academy.

propagation, nutrition, and regeneration were all fundamentally aspects of one single force, the *Bildungstrieb*.

Blumenbach explained that his experiments had led him to a new and significant finding, namely that regenerated parts were always *smaller* than the originals. When, in his subsequent medical practice, he observed the healing of a particularly extensive knee injury, he discerned that the restored tissue permanently manifested a depression, that is, was smaller in mass, and saw the parallel and inferred a generalization of very wide significance:

> That in all living creatures from man to mite and from cedar down to mold, a distinctive innate drive [*Trieb*] lies that remains actively at work over their entire lifespan, at the beginning to help them assume their definitive form, then to maintain it, and if that be destroyed, as much as possible to restore it.
>
> A drive (or propensity or endeavor, however one wishes to term it) that is entirely different both from the general properties of all bodies whatsoever and from the other characteristic forces [*Kräfte*] of organized bodies in particular, which appears to be one of the original causes of all propagation [*Generation*], nutrition and regeneration [*Reproduction*], and which I here, in order to avoid all misinterpretation and in order to distinguish it from other natural forces, provide the name formative drive [*Bildungs-Trieb*] (*nisus formativus*). (Blumenbach 1780b, 250–251)

The definition would remain virtually unaltered in all subsequent formulations, both in the texts devoted explicitly to the *Bildungstrieb* and in the texts that made use of the concept, for example, later editions of the *Handbuch der Naturgeschichte* and the various editions of his *Institutiones physiologicae* (Blumenbach 1787). He contended this new discovery should not be assimilated to older notions, such as *vis plastica*, associated with seventeenth-century discussion of organic life, or the more recent *vis essentialis* of C. F. Wolff. But Blumenbach had a more important argumentative task than distinguishing his position from other epigenesists. He had clearly broken with preformation theory and he needed to adduce his reasons for doing so, especially since he had in numerous earlier publications embraced the preformation theory of Haller.

Indeed, he swiftly saw the need to expand his argument to justify his apostasy. The result was his publication in 1781 of a book-length version: *Über den Bildungstrieb und das Zeugungsgeschäfte*. The core of the eighty-seven page book is the unaltered repetition of his article, but the supplements are worth attention. First, the book version opens with a long, clearly anxious kowtow to Haller, lamenting his "split from Haller—the man to whose writings and whose correspondence I owe so immeasurably much" (Blumenbach 1781, 6). Now he suggested there were problems in Haller's experimental results as summarized in his famous review of Wolff in the *Göttingische Gelehrten Anzeigen* (Haller 1760). Blumenbach confessed himself to have been convinced for a time, but now he raised reservations, claiming (in line with Wolff, incidentally, but without acknowledgment) that in the first days of embryonic development of the fertilized chicken egg there were no signs of blood vessels at all (Blumenbach 1781, 27–30).

In the book version, Blumenbach expanded as well on the difference of his view from earlier forms of epigenesis, *vis plastica* and *vis essentialis*. The weakness of the competing theories of epigenesis was "that all sorts of forces were assumed to carry out this process." He identified three forms of the epigenesis idea: outright animism ("The spiritualists have made the soul into the master builder"); Buffon's view (whereby an "inner model" present in the older organism provides the scheme after which "the basic material of the new organism is patterned"); and, finally, Caspar Friedrich Wolff's notion of a *vis essentialis* (Blumenbach 1781, 19). Blumenbach moved his discussion of the difference of his *Bildungstrieb* from Wolff's *vis essentialis* from a footnote in the original article into the body of the text, but the treatment amounted to no more than a citation from Wolff's *Theorie von der Generation* where Wolff gave a characterization of *vis essentialis* (Blumenbach 1781, 17–18). A great deal more would be required to settle the real differences between their views.

Having disposed of the rival theories, Blumenbach devoted the rest of his text to the role of the *Bildungstrieb* in propagation, nutrition, and regeneration, elaborating on the different regenerative capacities among animals, especially their limits in warm-blooded ones. In §35, Blumenbach brought to bear for the first time explicitly the experimental work of Joseph Gottlieb Kölreuter (1733–1806), demonstrating the mutation of one species of tobacco plant into another and then back over a series of generations, thus undermining the species essentialism that inspired preformation and eighteenth-century life science more generally (Kölreuter 1893). Kölreuter's results, Blumenbach averred, "ought to convert even the most committed defenders of the theory of evolution from their prejudice" (Blumenbach 1781, 61).

In 1782 Blumenbach issued the second, revised edition of his *Handbuch der Naturgeschichte*, introducing his theory of the *Bildungstrieb* in the revisions of Part II ("On Organized Bodies Generally") and of Part X ("On Plants"). Blumenbach's presentation of the idea of the *Bildungstrieb*, which he defined in essentially unchanged language, showed far greater confidence and concentration (Blumenbach 1782, 15). He gave preformation very short-shrift, based on its "numberless and irresolvable difficulties," and he did not even bother to distinguish his idea from other forms of epigenesis (Blumenbach 1782, 14). He argued that preformation had nothing to say concerning the ongoing organic processes of nutrition and regeneration, whereas his theory of the continued presence of the formative drive took all this directly into consideration (Blumenbach 1782, 15). Moreover, the regularity of species replication, and even the occurrence of birth defects and teratology generally, could be explained via the *Bildungstrieb* but not by preformation (Blumenbach 1782, 17–18). Finally, in the section on plants, Blumenbach offered a more extended discussion of Kölreuter's experiments, arguing that preformation was helpless to account for them, while they fit very nicely with his own theory of the *Bildungstrieb* (Blumenbach 1782, 471).

When he first presented his notion of the *Bildungstrieb* in 1780–1781, Blumenbach concentrated on how it answered certain physiological problems in organisms better than the alternative theories of preformation and of epigenesis. He did not dwell explicitly on the methodological or epistemological status of his concept. In the 1782 edition of his *Handbuch der Naturgeschichte*, his treatment of the idea of the *Bildungstrieb* once again gave no attention to this epistemological issue. He simply carried forward with his empirical exposition. But already in the third, 1788 edition of the *Handbuch*, he introduced a new section, immediately after defining his *Bildungstrieb*, with the following language:

> The *cause* of this formative drive can admittedly be so little adduced as that of attraction or gravity and other such generally recognized natural forces. It is enough that it is a distinctive force whose undeniable existence and broad influence throughout all of nature is revealed by experience, and whose constant phenomena offer a far more ready and clear insight into generation and many other of the most important topics of natural history than other theories offered for their explanation. (Blumenbach 1788, 14)

Note that the argument is presented in terms of the *general* order of nature: No strong distinction is made between the organic and the inorganic realms in terms of the nature of such forces, though, clearly, this particular force operated in organic forms. In its key features, the methodological argument parallels that developed by Reimarus in *Triebe der Thiere*. There is, as well, a tacit Newtonian analogy, without the mention of Newton by name.

In the second edition of his *Bildungstrieb* book (1789), Blumenbach became far more explicit about the Newtonian connection:

> The *term* formative drive, just like the *terms* attraction and gravity, etc. serves no more and no less than to denote a force whose constant effect is recognized but whose *cause* just as little as the causes of the other, nonetheless so generally recognized natural forces, remains for us a *qualitas occulta*. That does not hinder us in any way whatsoever, however, from attempting to investigate the effects of this force through empirical observations and to bring them under general laws. (Blumenbach 1789a, 32–33)

In the attached footnotes, Blumenbach referred directly to Newton, and then, in the context of the phrase *qualitas occulta*, to Voltaire's exposition of Newton, in particular to the passage where Voltaire argued that from a mere "blade of grass" to the order of the stars, *all* causes (physical as well as biological) were simply occult qualities (Blumenbach 1789a, 32–33n). This was standard epistemology of science in the wake of John Locke's discrimination of "nominal" from "real" essences, of empirical (external) observation from "inner" or ultimate reality of nature (Locke 1689). This epistemic modesty was common practice among all innovative natural scientists in the eighteenth century. Haller and Buffon had explicitly adopted it, and so had Caspar Friedrich Wolff (Roe 1979). As Peter

McLaughlin has argued, making the Newtonian appeal was constitutive for the experimental life sciences in the late eighteenth century (McLaughlin 1982). While it was epistemologically expedient, this may well have been disingenuous in many cases, for the forces were taken quite straightforwardly as real, even if the *ultimate* causes remained mysterious.

By 1788 Blumenbach had encountered significant resistance to his ideas—and from two fronts: the die-hard preformationists (Bonnet, Spallanzani, Caldani), but also the more aggressively naturalistic epigenesists—Thomas Sömmerring and Georg Forster and, of far greater importance, Caspar Friedrich Wolff (Roe 1981). One of Blumenbach's key arguments in 1781 had been that the *Bildungstrieb* encompassed and explained three vital functions—propagation, nutrition, and regeneration. In the 1789 version, nutrition gets scant attention. It is propagation and regeneration that Blumenbach offers in support of his theory in comparison with others. But it may also be that he had addressed the nutrition issue separately, in a prize-winning essay (1789) submitted to a competition sponsored by the Academy of Sciences in St. Petersburg, and presided over by his rival epigenesist, Caspar Friedrich Wolff (Blumenbach 1789b). While Wolff awarded Blumenbach a prize, he published a far lengthier work of his own on the topic, taking a sharply critical posture toward Blumenbach's views (Wolff 1789).

If we juxtapose the texts of 1781 and 1789 on the *Bildungstrieb*, Blumenbach by 1789 has dramatically improved the exposition of his *scientific* position. The structure of the argument is considerably clearer: After the historical background leading up to his own discovery, Blumenbach presents a thorough drubbing of the arguments for preformation, followed by a clear account of the advantages of his *Bildungstrieb* theory. He seems far more comfortable that he has made a major breakthrough and that he has defeated his rivals on that front.

But conceptually what exactly did *Bildungstrieb* signify, especially in relation to other natural forces? In the 1788 edition of his *Handbuch* Blumenbach found "throughout all nature the most unmistakable traces of a virtually general drive [!] to give matter a determinate form, which already in the inorganic realm is of striking effectiveness" (Blumenbach 1780a, 80). While he located his formative drive among other life-forces, that did not mean one could not draw *analogies* from the inorganic to the organic, for, he went on,

> even in the inorganic realm the traces of formative forces are so unmistakable and so general. Of formative *forces*—but not by far of the formative *drive* (*nisus formativus*) in the sense this term assumes in the current study, for it is a life-force [*Lebenskraft*] and accordingly as such inconceivable in inorganic creation—rather of other formative forces [*Bildungskräfte*], which provide the clearest proof in this inorganic realm of nature of determinate and everywhere regular formations [*Gestaltungen*] shaped out of a previously formless matter. (Blumenbach 1788, 80)

This is Blumenbach's most self-conscious distinction between the formative *forces* [*Kräfte*] of the inorganic realm and the formative *drive* (*Trieb*; note that it is always singular in Blumenbach's usage), unique to organic life. That makes it crucial for our consideration, but our first observation must be that Blumenbach classified the *Bildungstrieb* as a life-*force* [*Lebenskraft*] among others. We must still unearth the conceptual difference between *Kraft* and *Trieb*.

As Peter McLaughlin is quite right to maintain, Blumenbach did not do a very good job in explicating his *Bildungstrieb*: "what that is supposed to mean exactly is nowhere systematically elaborated" (McLaughlin 1982, 364). The Newtonian analogy did not minimize at all the actuality of the formative drive, but only denied access to its *ultimate* cause. This is a vital discrimination if we are to understand the relation between Blumenbach and Kant. McLaughlin offers three avenues to clarify the concept: first, how Blumenbach contrasted it with other theories and other forces; second, how he specified its typical laws of operation; and, finally, how he used it to explain other phenomena in natural history (McLaughlin 1982, 364). For McLaughlin, the contrast with C. F. Wolff is most illuminating, and the relation of the formative drive to organic matter is central. He observes: "Wolff's essential force [*vis essentialis*] was a *chemical* attraction-repulsion force" (McLaughlin 1982, 365). For Wolff, matter achieved various *levels* of organization, and once it passed a certain threshold, there ensued something of a chemical chain-reaction which initiated life. Wolff dedicated himself to experiments establishing the component constraints that directed the chain-reaction (Wolff 1764).

For Blumenbach, by contrast, McLaughlin believes the important question was the *inherent* relation between a distinctively organic matter and the forces unique to it. Blumenbach wished to distance himself from the notion of a continuity from the inorganic (in Wolff, the chemical) to the organic—that is, a materialist naturalism or "hylozoism." He found Wolff's notion of epigenesis problematic as much—or more—for the metaphysical quandaries as for the methodological ones. That is, all the earlier (materialist/naturalist) proponents of epigenesis sought to explain life as an emergent property arising out of matter itself. For La Mettrie, Buffon, and Holbach, according to McLaughlin, "life was a mechanical result of organization"—that is, of the general order of nature grounded in physics and chemistry—and this was what Blumenbach made of Wolff's *vis essentialis*. By contrast, Blumenbach aimed "to explain organic *form* through organic *matter*." That is, an *organic* force is "a force that only *has effect within* organic matter, not a force that somehow *causes* the transition from inorganic to organic matter." The *Bildungstrieb* did not explain life but rather presumed it. According to McLaughlin, Blumenbach's *Bildungstrieb* was not the *cause* of life but rather its *consequence* (McLaughlin 1982, 357, 359).[9]

How does this bear upon the key term, *Trieb*? Given what we have established both in Reimarus and in Blumenbach, we can reach two tentative judgments. First, epistemologically, an experimental life scientist could be content to establish (i.e., determinately recognize and categorize) organismic phenomena *without* being obliged to offer a causal *explanation*. Second, the organismic phenomena so recognized had the character of *immanent purposiveness*, which appeared beyond the power of mechanical explanation. These phenomena, both Reimarus and Blumenbach professed, seemed to be intrinsic (and ontologically specific) to *living* matter, not causal elements accounting for life as an emergent phenomenon within a unified natural order. An ontological mystery seemed acceptable for the sake of the determinacy of *empirical descriptions of organismic effects*.

[9] "The formative drive is not the *cause* of this leap [from inorganic to organic], but rather its *expression*" (McLaughlin 1982, 364) I share the view of Robert Richards (2002, 221ff.) that the relation in Blumenbach in fact tended to flow in the other direction, even if Blumenbach's metaphysical preferences inclined him to want to see it as McLaughlin reconstructs.

3 Blumenbach, Kant, and *Trieb*

The Preface to the second edition of Blumenbach's study on the *Bildungstrieb*, dated January 1789, advised readers that his earlier version was "immature" (Blumenbach 1789a, A4). What did Blumenbach intend by this statement, and by routine appeals in later versions of his *Handbuch* and of his dissertation on human variety as well, "not to confuse [this second edition] with the immature treatise that appeared under a similar title in 1781"? (Blumenbach 1789a, A4)[10] Timothy Lenoir famously claimed Blumenbach's "mature formulation resulted from his encounter with Kant's work" (Lenoir 1980, 84n). Lenoir argues that Blumenbach's "mature theory" was composed only "after he had begun to wrestle with Kant's philosophy of organic form," and ostensibly upon that basis (Lenoir 1980, 83). But I suggest that is historically quite problematic (Zammito 2012). We must regard Blumenbach's judgment of his earlier work in a more complex light. As we have noted, he was already making changes in his 1788 *Handbuch*, before we have any reason to suspect Kantian influence. Blumenbach's discomfort with the "immaturity" of his work of 1781 derives more from his tension with C. F. Wolff than from his affinity to Kant.

Blumenbach only began serious consideration of the philosophy of Kant at the end of the 1780s, as a consequence of the disputes surrounding Kant's reviews of Herder's *Ideen zur Philosophie der Geschichte der Menschheit*, and especially Kant's controversy with Georg Forster (Kant 1785, 1788; see Zammito 2011). Kant explicitly mentioned Blumenbach and his idea of the *Bildungstrieb* in his rejoinder to Forster in the *Teutscher Merkur* in January and February 1788 (Kant 1788, 180n). Blumenbach's first subsequent publication appeared almost immediately: the third edition of the *Handbuch der Naturgeschichte*, whose Preface was dated March 1788 (Blumenbach 1788). Unsurprisingly, it shows no evidence of Kantian ideas. Less than a year later, in January 1789, however, Blumenbach published his revised version of *Über den* Bildungstrieb and sent Kant a copy in acknowledgment of Kant's references to him in the 1788 essay (Kant 1790a). In this second edition of the *Bildungstrieb*,

[10] For later avowals along the same lines see, for instance, Blumenbach (1791, 13, 1797, 17n).

Blumenbach emphatically repudiated hylozoism, advancing beyond epistemological scruples to an ontological distinction between the general order of nature and the specific order of the organic: "No one could be more totally convinced by something than I am of the mighty abyss which nature has fixed [*befestigt*] between the living and the lifeless creation, between the organized and the unorganized creatures" (Blumenbach 1789a, 79). Indeed, Blumenbach shared Kant's skepticism about a bridge from the inorganic to the organic and about the phylogenetic continuity of life forms. What bound Blumenbach and Kant most together was their commitment to the fixity of species and their rejection of the reality of the "great chain of being." *Yet Blumenbach drew neither of these commitments from Kant.* They were already expressed with clarity in his dissertation of 1775 and especially the first edition of his *Handbuch* of 1779. These were basic issues for anyone taking up natural history in the eighteenth century.

McLaughlin identifies crucial changes that Blumenbach did introduce in 1791, after he had absorbed Kant's ideas not only from the 1788 essay but from the *Critique of Judgment* (1790), which Kant sent him. In his *Handbuch* of 1782, Blumenbach had written that "a particular, innate *drive*, active throughout its life, lies in every organized body" (Blumenbach 1782, 15). In the *Handbuch* of 1788, he wrote that one could find "throughout all nature the most unmistakable traces of a virtually general drive to give matter a determinate form, which already in the inorganic realm is of striking effectiveness" (Blumenbach 1788, 12–13). But in 1791 Blumenbach pruned the cited line as follows: one finds "in the entirety of organic nature the most unmistakable traces of a generally distributed drive to give matter a determinate form" (Blumenbach 1791, 14). The final clause from 1788 was eliminated altogether. In 1789, Blumenbach compared the *Bildungstrieb* to "the *terms* attraction and gravity [...] generally recognized natural forces [...]" (Blumenbach 1789a, 32–33). But in 1797 he changed this to: "The term Bildungstrieb just like all other life forces such as sensibility and irritability [...]" (Blumenbach 1797, 18). The thrust, as McLaughlin notes, was to make a radical distinction between the organic and the inorganic realms and to explain the drive exclusively in terms of the former.

In 1799, Blumenbach inserted that this life-force was "distinguished from all merely mechanical formative force [...] (by the connection of merely mechanical to the purposively modifiable elements in this drive)" (Blumenbach, 1799, 17). This was clearly an incorporation of Kantian language, for Kant had complimented Blumenbach precisely for this in his letter of August 1790, a characterization of his achievement that Blumenbach found quite flattering. He elaborated this idea in terms of a notion of organic form that also showed Kantian overtones: "by virtue of its capacity to bring to form all the endlessly multifarious different determination of organized bodies and their parts the complexly organizable generative matter in even so multifarious as thoroughly purposive modifications into determinate forms and thus [...] initially at conception causes this gradual development [...]" (Blumenbach 1799, 17). This had been what Kant found most gratifying in Blumenbach's new book of 1789, as he reported in his letter of acknowledgment to Blumenbach (Kant 1790a, 176–177). In the *Critique of Judgment* he elaborated:

> Blumenbach [...] rightly declares it to be contrary to reason that raw matter should originally have formed itself in accordance with mechanical laws, that life should have arisen from the nature of the lifeless, and that matter should have been able to assemble itself into the form of a self-preserving purposiveness by itself; at the same time, however, he leaves natural mechanism an indeterminable but at the same time also unmistakable role under this inscrutable *principle* of an original *organization*, on account of which he calls the faculty in the matter in an organized body (in distinction from the merely mechanical *formative force* [Bildungskraft] that is present in all matter) a *formative drive* [*Bildungstrieb*] (standing, as it were, under the guidance and direction of that former principle). (Kant 1790b, 424; my own translation)

Thus, Kant picked up and underscored precisely the *metaphysical* implications of Blumenbach's methodological replacement of *Kraft* by *Trieb*, and thrust the problem of life beyond the grasp of empirical science into an inscrutable principle of an original organization. That ruptured not only the unity of actual nature, of which life forms were incontestably an inherent part, but also the possibility of an empirical science of life. In response, the philosophical as well as the empirical research communities of the next generation in Germany would accept neither of these interdicts. *Naturphilosophie* and empirical life science asserted themselves *against* Kant and Blumenbach by taking *Trieb* as an essential concept of nature and of spirit, indispensable for natural science and for philosophy.

References

Blumenbach, Johann Friedrich. 1779. *Handbuch der Naturgeschichte*. Göttingen: Dieterich.

———. 1780a. *Handbuch der Naturgeschichte. Zweyter Theil.* Göttingen: Dieterich (Continuously Paginated, with an Appended, Unpaginated Register for both Volumes; All Subsequent Editions Would be Bound as a Single Volume).

———. 1780b. Über den Bildungstrieb (*Nisus formativus*) und seinen Einfluß auf die Generation und Reproduction. *Göttingisches Magazin der Wissenschaften und Litteratur* I (5): 247–266.

———. 1780c. Von der Federbusch-Polypen in den Göttingener Gewässern. *Göttingisches Magazin der Wissenschaften und Litteratur* I (4): 117–127.

———. 1781. *Über den Bildungstrieb und das Zeugungsgeschäfte*. Göttingen: Dieterich (reprinted: Stuttgart: G. Fischer, 1971).

———. 1782. *Handbuch der Naturgeschichte*. 2nd ed. Göttingen: Dieterich.

———. 1787. *Institutiones physiologicae*. Göttingen: Dieterich (English translation: Blumenbach, Johann Friedrich. 1817. *Institutions of Physiology*. London: Bensley and Son).

———. 1788. *Handbuch der Naturgeschichte*. 3rd ed. Göttingen: Dieterich.

———. 1789a. *Über den Bildungstrieb*. 2nd ed. Göttingen: Dieterich.

———. 1789b. Versuch einer Beantwortung der von der kaiserlichen Akademie der Wissenschaften zu St. Petersburg, zum drittenmal aufgegebenen Preisfrage, *uti nutritio aequabilis, etc.* In *Zwo Abhandlungen über die Nutritionskraft welche von der Kayserlichen Academie der Wiſsenschaften in St. Petersburg den Preis getheilt haben. Die erste von Herrn Hofrath Blumenbach, die zwote von Herrn Prof. Born. Nebst einer fernern Erläuterung eben derselben Materie von C. F. Wolff*, 7–16. St. Petersburg: Royal Academy of Sciences.

———. 1791. *Handbuch der Naturgeschichte*. 4th ed. Göttingen: Dieterich.

———. 1797. *Handbuch der Naturgeschichte*. Götttingen: Dieterich.

———. 1799. *Handbuch der Naturgeschichte*. Götttingen: Dieterich.

Broberg, Gunnar. 1994. *Homo sapiens*: Linnaeus's Classification of Man. In *Linnaeus: The Man and His Work*, ed. Tore Frängsmyr and Sten Lindroth, 156–194. Canton, MA, USA: Science History Publications/USA.

Condillac, Abbé de. 2004. *Traité des animaux*. Edited by Michel Malherbe. Paris: Vrin.

Engels, Eve-Marie. 1994. Die Lebenskraft—metaphysisches Konstrukt oder methodologisches Instrument? Überlegungen zum Status von Lebenskräften in Biologie und Medizin im Deutschland des 18. Jahrhunderts. In *Philosophie des Organischen in der Goethezeit*, ed. Kai Torsten Kanz, 127–152. Stuttgart: Steiner.

Haller, Albrecht von. 1936 [1755]. *Dissertation on the Irritable and Sensible Parts of Animals*. Translated by S. A. D. Tissot, edited by O. Temkin. Baltimore: Johns Hopkins University Press.

———. 1760. *Review of Wolff's Theoria generationis. Göttingische Gelehrten Anzeigen* 143 (29. November 1760): 1226–1232.

Jaynes, Julian, and William Woodward. 1974a. In the Shadow of the Enlightenment: I. Reimarus against the Epicureans. *Journal of the History of the Behavioral Sciences* 10: 3–15.

———. 1974b. In the Shadow of the Enlightenment: II. Reimarus and His Theory of Drives. *Journal of the History of the Behavioral Sciences* 10: 144–159.

Kant, Immanuel. 1785. *Recension von J. G. Herders Ideen zur Philosophie der Geschichte der Menschheit* (1785). Vol 8, 43–66. Berlin: Akademie der Wissenschaften, now de Gruyter [henceforth AA].

Kant. 1788. Über den Gebrauch teleologischer Principien in der Philosophie, AA 8, 157–184.

———. 1790a. [Letter to Blumenbach, August 5,1790], AA 11, 176–7.

———. 1790b. *Kritik der Urtheilskraft*, AA 5, 163–486.

Kästner, Abraham. 1774. Verbatim Report of Blumenbach's Presentation to the Academy. *Göttingische Gelehrten Anzeigen* (1. October 1774), II: 1009–1011.

Kölreuter, Joseph Gottlieb. 1893. *Vorläufige Nachricht von einigen das Geschlecht der Pflanzen betreffenden Versuchen und Beobachtungen, nebst Fortsetzungen 1, 2 und 3*. Leipzig: Engelmann.

Lenhoff, Sylvia, and Howard Lenhoff. 1986. *Hydra and the Birth of Experimental Biology—1744. Abraham Trembley's Memoirs Concerning the Natural History of a Type of Freshwater Polyp with Arms Shaped Like Horns*. Pacific Grove, CA: Boxwood.

Lenoir, Timothy. 1980. Kant, Blumenbach, and Vital Materialism in German Biology. *Isis* 71: 77–108.

Linnaeus, Karl. 1758. *Systema naturae*. 10th ed. Holmiae.
Locke, John. 1689. *Essay Concerning Human Understanding*. London: Routledge.
Mayr, Ernst. 1982. Geleitwort. Reimarus 1760, 10–14.
McLaughlin, Peter. 1982. Blumenbach und der Bildungstrieb: Zum Verhältnis von epigenetischer Embryologie und typologischem Artbegriff. *Medizinhistorisches Journal* 17: 357–372.
Meier, Georg Friedrich. 1742. *Beweiss: daß keine Materie denken könne*. Halle: Hemmerde.
———. 1747. *Rettung der Ehre der Vernunft wider die Freygeister*. Halle: Hemmerde.
———. 1749. *Versuche eines neuen Lehrgebäudes von den Seelen der Thiere*. Halle: Hemmerde.
Reimarus, Hermann Samuel. 1725. *Instinctum Brutorum Existentis Dei Ejusdemque Sapientissimi Indicem*, Reprinted and Translated into German in the Critical Edition of Reimarus, 1760, 757–780.
———. 1985 [1754]. *Die vornehmsten Wahrheiten der natürlichen Religion*. Vol. 2. Edited by Günter Gawlick. Göttingen: Vandenhoeck & Ruprecht
———. 1982 [1760]. *Allgemeine Betrachtungen über die Triebe der Thiere, hauptsächlich über ihre Kunsttriebe*. Edited by Jürgen von Kempski. Göttingen: Vandenhoeck & Ruprecht.
Richards, Robert. 2002. *The Romantic Conception of Life*. Chicago: University of Chicago Press.
Roe, Shirley. 1979. Rationalism and Embryology: Caspar Friedrich Wolff's Theory of Epigenesis. *Journal of the History of Biology* 12: 1–43.
———. 1981. *Matter, Life and Generation: Eighteenth-century Embryology and the Haller-Wolff Debate*. Cambridge: Cambridge University Press.
Wolff, Caspar Friedrich. 1764. *Theorie von der Generation in zwei Abhandlungen erklärt und bewiesen (1764). Theoria generationis (1759)*. Berlin. Birnstiel. (Reprint: Hildesheim: Olms, 1966.)
———. 1789. Von der eigenthümlichen und wesentlichen Kraft der vegetabilischen sowohl als auch der animalischen Substanz. In *Zwo Abhandlungen über die Nutritionskraft welche von der Kayserlichen Academie der Wißenschaften in St. Petersburg den Preis getheilt haben. Die erste von Herrn Hofrath Blumenbach, die zwote von Herrn Prof. Born. Nebst einer fernen Erläuterung eben derselben Materie von C. F. Wolff*. St. Petersburg: Royal Academy of Sciences.
Wundt, Max. 1939. *Die deutsche Schulmetaphysik des 17. Jahrhunderts*. Tübingen: Mohr Siebeck.

Zammito, John H. 2011. History of Philosophy vs. History of Science: Blindness and Insight of Vantage Points on the Kant-Forster Controversy. In *Klopffechtereien—Missverständnisse—Widerspruche?. Methodische und methodologische Perspektiven auf die Kant-Forster-Kontroverse*, ed. Rainer Godel and Gideon Stiening, 225–244. Paderhorn: Fink.

———. 2012. The Lenoir Thesis Revisited: Blumenbach and Kant. *Studies in History and Philosophy of Science*, Part C: *Studies in History and Philosophy of Biological and Biomedical Sciences* 43: 120–132.

4

Stoic Dispositional Innatism and Herder's Concept of Force

Nigel DeSouza

The concept of drive or *Trieb* in Herder is not ontologically basic. This does not mean that it is not of basic importance to Herder, and one may profitably explore how Herder makes use of this concept in his philosophy, above all in his philosophy of life.[1] What interests me here, however, is something that *is* ontologically basic, and out of which drives emerge, namely, force or *Kraft*. From Herder's earliest writings we know that the triad of concepts of space (*Raum*), time (*Zeit*), and force (*Kraft*) were fundamental to his metaphysics as he sought to critically engage with and move beyond the Wolffian-Baumgartenian philosophy or *Schulphilosophie* he had been taught by Kant at the University of Königsberg in the early 1760s.[2] I have explored elsewhere the origins of Herder's conception of force in his critical engagement with this tradition, Leibniz and, above

[1] See, for example, the discussion in Pénisson (2002), DeSouza (2012a), and Zammito (2017).

[2] See, for example, *Versuch über das Sein* (1763) in Herder (1985) and *Grundsätze der Philosophie* (1769) in Herder (1987).

N. DeSouza (✉)
University of Ottawa, Ottawa, ON, Canada
e-mail: ndesouza@uottawa.ca

M. Kisner, J. Noller (eds.), *The Concept of Drive in Classical German Philosophy*,
https://doi.org/10.1007/978-3-030-84160-7_4

all, the pre-critical Kant, a conception that is pivotal to his interactionist theory of the soul-body relationship and that comes to its fullest and broadest exposition in the theory of organic forces deployed in both the *Ideen zur Philosophie der Geschichte der Menschheit* (1784–1791) and *Gott: einige Gespräche* (1787).[3] In this paper, I wish to turn to another influence on Herder's concept of force, one that I think is key for understanding his concept of drive. The most important and immediate sources of this influence, as we shall see, are Leibniz and Shaftesbury, but that is because they transmitted and importantly modified it. The influence I am speaking of is that of the philosophy of the Stoa,[4] and, specifically, of their dispositional innatism. As my topic is the origins of Herder's concept of force, I will focus on Herder's earliest, foundational writings from the 1760s, and will consider, in turn, Leibniz, Shaftesbury, and finally Baumgarten, whose aesthetic project served as an ideal springboard, or rather foil, for Herder's own proposal for an aesthetics that truly descended to the "Grund der Seele".

1 Leibniz

There is no better place to begin than with these words of Leibniz:

> Ist die Seele eine leere Tafel, nach Aristoteles und Locke; oder hält sie die Gründe von vielen Begriffen und Lehrsätzen ursprünglich in sich, die die äußern Gegenstände nur bei Gelegenheit aufwecken? Diese Urbegriffe nennen die Stoiker *notiones communes*, *prolepses*, κοινὰς ἐννοίας; Julius Scaliger *semina aeternitatis*, *Zopyra*, lebendiges Feuer, leuchtende Züge, in uns verborgen, die der Gebrauch der Sinne aufweckt, wie Funken, die der Schlag aus dem Gewehr treibt. (Herder 1987, 33)

[3] See, for example, DeSouza (2012b, 2016a, b, and 2018).

[4] Recently there has been important work done on the Stoic background to Herder's philosophy; the present paper seeks to add to that work in a complementary fashion, bringing out a different aspect of Stoic philosophy important for understanding Herder's philosophy. See Piirimäe (2020) and Heinz (2020).

Only, these are not *exactly* Leibniz's words. They are Herder's translation and partial abbreviation of Leibniz's words from the beginning of the Preface to his *Nouveaux essais sur l'entendement humain* (1765).[5] Herder's friend and mentor, Hamann, had been the first to inform him in January 1765 of the recent publication of a volume of works by Leibniz, edited by R. E. Raspe, in which Leibniz's critical engagement with Locke's *Essay Concerning Human Understanding* (1690) took pride of place. Despite Hamann's dismissive remarks about the *Nouveaux essais* ("sein scholastisches Geschwätz ist mir niemals recht nach meinem Geschmack gewesen"), Herder's biographer, Rudolf Haym, claims that Herder immediately set about assimilating its contents through extensive excerpts, entitled *Wahrheiten aus Leibniz.*[6] Indeed, it is well-established that Herder was profoundly influenced by Leibniz, and that more than once he referred to a trio of thinkers who were particularly important in shaping his thought: Leibniz, Spinoza, and Shaftesbury.[7] However, it is still a matter of scholarly debate how early Herder in fact engaged with the writings of Leibniz. Beate Dreike and others maintain that Herder only really began to do so

[5] Here is the corresponding passage from Leibniz, *Nouveaux essais*, p. 4, in the edition by Raspe (1765) that Herder read: "Il s'agit de savoir si l'Ame en elle-même est vuide entierement comme des tablettes, où l'on n'a encore rien écrit (*tabula rasa*) suivant Aristote & l'Auteur de l'Essai, & si tout ce, qui y est tracé, vient uniquement des sens & de l'experience? Ou si l'Ame contient originairement les principes de plusieurs notions & doctrines, que les objets externes reveillent seulement dans les occasions, comme je le crois avec Platon & même avec l'Ecole & avec tous ceux, qui prennent dans cette signification le passage de S. Paul (Rom. II. 15.) où il marque, que la Loy de Dieu est écrite dans les cœurs? Les Stoïciens appelloient ces principes *notions communes*, *Prolepses*, c'est à dire des assumtions fondamentales, ou ce, qu'on prend pour accordé par avance. Les Mathematiciens les appellent *notions communes* (κοινὰζ εννοἰαζ). Les Philosophes modernes leur donnent d'autres beaux noms, & Jules Scaliger particulierement les nommoit *Semina aeternitatis*, item *Zopyra*, comme voulant dire des feux vivans, des traits lumineux, cachés au dedans de nous, que la rencontre des sens & des objets externes fait paroitre comme des etincelles, que le choc fait sortir du fusil; & ce n'est pas sans raison, qu'on croit, que ces eclats marquent quelque chose de divin & d'éternel, qui paroit sur tout dans les verités necessaires."

[6] Herder (1846, 299), (Haym 1885, 265). Haym maintains that the title given to Herder's excerpts is "unpassend", but this is not necessarily true, given that Herder not only abbreviates Leibniz's words, but also adds words of his own, and, in one case at least, deliberately omits certain topics, for example, studiously avoiding in his excerpts each mention Leibniz makes of his theory of the pre-established harmony of soul and body, a theory that Herder would subject to fundamental critique in his contemporaneous text *Ueber Leibnizens Grundsätze von der Natur und Gnade*. It should also be noted that the published versions of Herder's *Wahrheiten* are incomplete: They only extend to Chapter 1 of Book II, whereas the manuscript in fact reaches Chapter 27 of Book II. For discussion of *Wahrheiten*, see DeSouza (2016a, 105–108).

[7] See, for example, Dreike (1973).

in 1769, which corresponds to the date on the manuscript of *Wahrheiten*. It is not my intention to resolve this debate here. While I do think that what I will discuss below points to the influence of the *Nouveaux essais* on his thought earlier than 1769 and thus sides with Haym's view, my intention here is limited to showing how, through his study of two of his avowed greatest intellectual influences, Herder's philosophy in its earliest formative phase of the 1760s was both influenced by and creatively inserted itself into the tradition of Stoicism in European thought.

These important biblio-biographical details considered, let us return to the quote. The first point to note is that this is the first, and perhaps *only*, explicit mention in Herder's writings of the Stoic concepts of *prolepsis* and *common notion* (*koine ennoia*), the concepts that, I wish to maintain, directly or indirectly, play a role in his, at this time, developing understanding of *Kraft*. The second point to note is the role these concepts are playing in Leibniz's text: They are deployed in order to add historical weight to the epistemological tradition that Leibniz is seeking to oppose to that of Locke. One naturally thinks here of rationalism versus empiricism, but a more accurate term for the main position Leibniz is seeking to argue for (since he defends others, such as implicit knowledge[8]) wouldbe "dispositional innatism".[9] Indeed, throughout the Preface and Book I, entitled "Des notions innées", Leibniz returns, again and again, to the defence of this position. In invoking, among others, the Stoics and the terms *prolepsis* and *koine ennoia*, Leibniz is implying that the Stoics themselves were dispositional innatists. On this point there has been lively debate among scholars for over a century as to whether the Stoics were empiricists or innatists,[10] but more recently it has come to be accepted as the "standard view" that among both the early and the late Stoa "the Stoic account of concept acquisition amounts to a form of dispositional innatism" (Klein 2011, 115). Dominic Scott, in an article entitled "Innatism and the Stoa", seeks to reconcile the conflicting passages among both early and late Stoic authors by distinguishing between adventitious concepts acquired via the senses (such as hot and cold, black and white) and

[8] See, for example, Leibniz (1765, 76–77).

[9] See, for example, Bui (2017), Scott (1988).

[10] For a brief summary, see Scott (1988, 150). For detailed discussion, see Dyson (2009).

moral concepts that have an internal source in the human being (Scott 1988, 146). Scott also importantly distinguishes between psychological and epistemological innatism, the former involving a description of how concepts and beliefs naturally arise in individuals, while the latter justifies those concepts and beliefs as trustworthy, for merely showing that and how concepts are acquired need not entail that they should be believed (Scott 1988, 136, 146). The dispositional innatism that Scott sees in the Stoics primarily involves moral notions, which are taken to be *emphutoi* or innate not because any concepts are innately fully known, but rather because individuals possess a disposition or inclination towards virtuous behaviour that in turn implies a disposition to form the (ethical) belief that that behaviour is right. It is in this sense that certain ethical ideas and beliefs can be taken to be innate (Scott 1988, 144–145), a position associated with, among others, Chrysippus, Diogenes, Epictetus, and Seneca, who, famously, said: "Nature could not teach us this [sc. the good]; she has given us the seeds of knowledge, but not the knowledge itself".[11] This particular quote is instructive because it not only defends psychological innatism, according to which certain knowledge arises in us connately or naturally, it also points to the widely shared grounds among the Stoa for subscribing to epistemological innatism, in a word, that nature is providential and thus to be trusted (cf. Scott 1988, 147).

Now in the case of Leibniz in the *Nouveaux essais*, while he too will speak of innate moral notions, he does not limit himself to the moral. Herder faithfully reproduces Leibniz's oft-repeated claim that while the senses are needed for our actual knowledge, they are not the source of general truths in ethics, mathematics, logic, and metaphysics, which, along with their universal necessity, must be found in the soul and brought to awareness by attention, not simply read "*à livre ouvert*", in Herder's words (Herder 1987, 33; Leibniz 1765, 5). Leibniz's elaboration of how such truths are in the soul or mind (*l'esprit*) is also carefully noted by Herder:

[11] "Hoc [bonum] natura nos docere non potuit; semina nobis scientiae dedit, scientiam non dedit." Seneca, *Epistulae Morales ad Lucilium*, 120, quoted in Scott (1988, 140).

> Wir sind uns selbst gleichsam eingeboren. Wesen, Einheit, Substanz, Dauer, Veränderung, Handlung, Vorstellung, Vergnügen und tausend andre Gegenstände von unsern Intellektual-Ideen, die unsrem Verstande unmittelbar gegenwärtig sind, obgleich nicht immer bemerkt. Das ist ein Marmor, der nicht leere Tafel ist, sondern in dem die Adern zu einer Figur des Herkules schon inliegen: sie also bloß entdecken, glätten, und das Hindernde abtun. So sind uns die Ideen und Wahrheiten eingeboren [*innées*], als Neigungen, Disposition, Habituden [*virtualités*], aber doch daß die Vermögenheiten [*virtualités*] immer von Handlungen begleitet sind, die zwar unmerklich sind, ihnen doch aber zusagen. (Herder 1987, 34; Leibniz 1765, 7)

Leibniz thus not only reflects the same dispositional innatism as the Stoics, he also elaborates on it in ways that were influential for the young Herder, who assimilates, as we shall see, the idea that the soul already contains basic ideas and truths that are first expressed in inclinations, dispositions, and drives. Agreeing with Leibniz against Plato's theory of recollection, Herder repeats that there must have been some first state in which innate principles and truths were present in the (both passive *and* active) mind (*Geist*), and that these can and must unfold (*sich entwickeln*) in the course of our lives (Herder 1987, 41; Leibniz 1765, 35, 36). While we start in actual knowledge from particular truths and composite ideas, these all are ultimately based on general principles which (Herder adding the words in italics), "*machen da Leben und Bindung aus*, sie sind da das, was Muskeln und *Nerven* [*tendons*] jeden Augenblick im Gehen sind, ohne daß wirs merken. Der Geist stützt sich auf sie immer, aber denkt an sie nicht deutlich und abgesondert, weil das große Aufmerksamkeit fodert" (Herder 1987, 41; Leibniz 1765, 40). Herder seems here to be linking the function of these innate principles and truths to the operation of life itself, a connection that becomes central in his later writings.

Indeed, Herder finds more support for this association when Leibniz turns to the question of innate practical principles and associates the maxim of pursuing joy and avoiding sorrow with an *instinct* that is not known only through reason or the natural light, but is instead individually initially based on what is only sensed, on "innerliche confuse Erfahrungen" (Herder 1987, 42; Leibniz 1765, 45). Leibniz here echoes the ideas we encountered above in the Stoics when he argues against

Locke's claim that the desire for happiness and aversion to misery, qua *inclinations*, are not products of reason and thus not practical truths. Rather, Leibniz responds, "[o]r le penchant, exprimé par l'entendement, passe en precepte ou verité de practique: & si le penchant est inné, la verité l'est aussi, n'y ayant rien dans l'ame qui ne soit exprimé dans l'entendement" (Herder 1987, 43; Leibniz 1765, 47). That is, if the disposition or inclination is innate, so is the precept or truth to which it gives rise. This operation of the instinct, Leibniz adds, is not restricted to practice, but in fact extends to theoretical truths as well (in Herder's translation): "das sind die ersten Grundsätze der Wissenschaft, die wir ohne Grund, durch einen natürlichen Trieb [*instinct*] annehmen" (Herder 1987, 43; Leibniz 1765, 47). The natural light can thus work in tandem with instinct: "[e]s gibt auch Principien des Instinkts, die Conclusionen des natürlichen Lichts sind: so Menschlichkeit.—Der gemeine Mensch braucht sie nach Instinkt, der Logiker entkleidet sie" (Herder 1987, 43; Leibniz 1765, 48). As Leibniz also notes, we act humanely because it pleases us, and also because reason (eventually) shows us it is right. Indeed, because of the sheer importance of morality, Leibniz continues and Herder notes, God has given human beings instincts to lead them directly to right action without the aid of reasoning, as expressed in natural instincts such as "[un] instinct general de *societé*", love between the sexes and for offspring (extending even to affection and gentleness between animals), even if reasoning is nevertheless ultimately required for morality to be made completely certain (Herder 1987, 43; Leibniz 1765, 49–50). Summarizing these complementary roles of the instinct/feeling and reason, Leibniz writes:

> tout sentiment est la perception d'une verité, & que le sentiment naturel l'est d'une verité innée, mais bien souvent confuse, comme sont les experiences des sens externes: ainsi on peut distinguer les *verités innées* d'avec la *lumiere naturelle* (qui ne contient que de distinctement connoissables) comme le genre doit étre distingué de son espece, puisque *les verités innées* comprennent tant *les instincts* que *la lumiere naturelle* (Leibniz 1765, 50–51)

Herder's abbreviated version of this passage is as follows: "Ohne Vernunft zwar freilich nicht überzeugend; aber doch eine Empfindung, ein dunkles

Gefühl einer eingebornen Wahrheit, ist, die aufgeklärt werden muß" (Herder 1987, 44).

It is perhaps no accident that Herder ostensibly inaccurately, but deliberately, translates Leibniz's French word "confuse" with the German "dunkel", which is normally the translation for "obscure". Herder was closely acquainted with Leibniz's *Meditationes de cognitiones, veritate, et ideis* (1684)[12] in which Leibniz clearly distinguishes obscure (*obscura*) knowledge from clear knowledge, the latter alone being, in turn, confused or distinct. And over the course of several short philosophical writings from 1766 to 1769, Herder turns time and again to what he calls "die dunkelsten Gegenden der Seele", which are misunderstood, misrepresented, or just plain unexplained by philosophers such as Wolff and Baumgarten.[13] What we see here, to speak generally, is Herder's own attempt to realize the transformation of philosophy into anthropology that he calls for as early as 1765, in which the so-called *Aufwertung der Sinnlichkeit* plays a fundamental role. Indeed in his first philosophical piece, the *Versuch über das Sein* (1763), Herder rejects all attempts at logical proofs of the concept of being in favour of an approach that understands it to be the most sensuous (*allersinnlichste*) and thus most certain, and most unanalysable (*unzergliederlichste*), concept, that is primordially felt, not rationally known (Herder 1985). Similarly, the most obscure regions of the soul are understood to be "pregnant" with meaning that is first felt and in which all knowledge and culture is ultimately rooted.[14] A full account of Herder's views during this formative period (which I and others have explored in detail elsewhere[15]) is neither my object nor purpose here. Rather, what I want to bring out is a hitherto neglected dimension of Herder's early philosophy, and, in particular, of his concept of *Kraft*. If we now bring together this idea of the obscure or *dunkel* and the

[12] In Herder's Nachlass, there is a page of excerpts from this work dated 1768, see Herder (1979, 198).

[13] See, for example, *Über Christian Wolfs Schriften*, Herder (1987, 10), where Herder says of Wolff, "[d]ie dunkelsten Gegenden der Seele [...] sind von ihm unbeleuchtet geblieben" (for discussion, see DeSouza 2016b); and Herder's critical engagement with Baumgarten's *Aesthetica*, considered below in Sect. 3, which takes up the study of the "dunkel".

[14] Hence the title of Hans Adler's book, *Die Prägnanz des Dunklen: Gnoseologie, Ästhetik, Geschichtsphilosophie bei Johann Gottfried Herder* (1990).

[15] See Heinz (1994), DeSouza (2016a, 2017).

concept of *Kraft*, we can begin to see how Herder's ideas reflect the Stoic ideas of prolepsis and *koinai ennoia* that we have already seen reflected in Leibniz's *Nouveaux essais*.

In a text from 1766 to 1769[16] entitled *Plato sagte: daß unser Lernen bloß Erinnerung sei*, Herder begins as follows: "Die Seele tritt auf die Welt: Vorstellungskraft ist ihr Wesen: aber sie ist *sich selbst* ganz ihr Gedanke—der dunkle, aber lebhafteste Begriff ihres Seyns erfüllt sie ganz [...] Dieser Gedanke ist ein dunkles, aber Einziges lebhaftes Gefühl: so stark und fruchtbar, daß alle übrige künftige, auch sinnliche, und noch mehr Abstrakte Begriffe in ihm liegen" (Herder 1994, 175). Herder begins from the notion of the soul as a representational force/power, familiar from Wolff's philosophy, and makes the epistemological starting point an obscure thought the soul has of its own being that, when it first reaches the level of awareness, is *felt* in the most lively way. In a manner analogous to the innate principles and truths that Leibniz defended above, the soul at this stage still contains its future concepts. In the passages that follow, as in other contemporaneous pieces (in which he is moving well *beyond* Leibniz), Herder explains how the soul realizes itself in space, time, and through force by building itself a body through whose senses it comes to engage with the structurally similar world that itself exists in space, time, and through relations of force. Herder draws structural analogies between the sense of sight and space (*nebeneinander*), hearing and time (*nacheinander*), and force and touch (*durcheinander*), and it is through this physical embodiment that the soul unfolds and develops its being, in the process gradually coming to awareness of the concepts of being, space, time, force, shape, extension, size, and so on (Herder 1994, 178). As with the dispositional innatism of the Stoics and Leibniz, the soul is *inclined* to unfold and develop these concepts; but with Herder, this unfolding of the soul, of the *Seelenkräfte*, is joined to a conception of life whereby these forces become *organic* forces. There are, in this process by which the soul builds itself a body and realizes itself in the world, at least three different dimensions to Herder's concept of force

[16] See Heinz (1994, 43). While more research would be needed to adequately ground this claim, if we take the text of *Plato sagte* to have been composed earlier than 1769, what I present here corroborates Haym's claim that Herder studied Leibniz's *Nouveaux essais* earlier than 1769, since my analysis points to echoes of Leibniz's text in *Plato sagte*.

at play. First, there is the physical-mechanical dimension according to which the soul is able to build itself a body because the soul-force is able to harness the forces of attraction and repulsion in matter by acting on the inner principle of a physical monad, this interaction being rendered possible by the ultimate ontological identity of soul and body *qua* force (here one sees the influence of Newton and the pre-critical Kant on Herder). Second, there is the Aristotelian-Leibnizian dimension of the soul as an organic force *qua* principle of life. And third, and closely related to the second, is the Stoic dimension I am proposing here of the soul with its thought-force (*Gedankenkraft*) and concept of being as stocked with tendencies, inclinations, drives that give rise to and undergird all human behaviour and knowledge.

2 Shaftesbury

Now there is one crucial dimension to this Stoic influence on Herder's conception of the human soul that needs to be filled in, namely, the moral, and with it we also add to our study the other philosopher who exercised great influence on Herder: Shaftesbury. For in the writings of Herder from the 1760s that we have been considering, Herder speaks of "Geschlechtertrieb [als] *Instinkt*", "Gesellschaftstriebe im Menschen [als] natürlich", and so on (Herder 1987, 55). Indeed, as early as 1765, in addressing the topic of how philosophy can be made more useful to common people (*das Volk*), Herder rejects the moral philosophy of his day with its science of duties and obligations, asking:

> Hat das Volk wirkliche praktische üble *Grundsätze* wider die Moralité: ich meine nein! denn der meiste Teil von ihnen handelt in der Tat nach keinen Grundsätzen im strengsten Verstande. *Wornach[?]* der Zaum, der ihn leitet—Dank der Natur! die uns schuf—sind nicht Kenntnisse, sondern Empfindungen, und diese sind alle gut: sie sind Stimmen des *Gewissens*, unsres Führers, von Gott gesandt; sie sind schwächer zu machen, aber nicht zu verdunkeln. Ich schließe weiter. Alles, was die *Grundsätze* und Maximen der Moral sagen, weiß ein jeder, eingewickelt und dunkel. Man zeige mir eine Regel der Moral, die ich als Mensch, nicht weiß; das beste Kennzeichen: sie kommt mir nicht zu. Aber *dunkel*? ja allerdings dunkel;

> aber die Dunkelheit ist ein Schatte ihrer Würde, sie ist unzertrennlich von dem, was rührend ist. All Licht, was ihr der Philosoph gibt, macht eine Sache deutlich, die mir vorher gewiß war; er lehrt sie meinem *Verstande,*— und mein Herz, nicht der Verstand muß sie fühlen. (Herder 1985, 115–116)

To be sure, Herder's emphasis of the morally motivating and cognitive power of the obscure/*dunkel*, his notion of "Gesellschaftstriebe" above, the dual roles of feeling and reason, and the characterization of these sentiments as gifts of Nature/God recall the ideas we highlighted earlier in *Wahrheiten aus Leibniz.* But these equally betray the deep influence of Shaftesbury. In fact, in a letter to Kant from November 1768, Herder not only mentions several of Shaftesbury's works by name and praises him for having "so viel Menschliche Weltweisheit", he also calls him "den Freund unsers alten Leibnitz, dem dieser ungemein viel schuldig ist, u. den er sehr gerne gelesen" (Herder 1984, 115).[17] With respect to the above quote, Herder can be interpreted as combining the Leibnizian notion of obscure knowledge with Shaftesbury's moral sense theory. For although Leibniz also speaks of an "instinct" and "sentiment" in this regard, as we saw above, Shaftesbury's framework and analysis is far richer. Now my working hypothesis throughout has been that the Stoic notion of prolepsis, directly or indirectly, contributed to Herder's conception of the obscure or *dunkel* with respect to human knowledge and action. In Shaftesbury, whose *Characteristicks of Men, Manners, Opinions, Times* (1711) he began to study in the mid-1760s,[18] Herder would have read in Shaftesbury his explicit attestations to the debts he owed to the Stoics as well as his many discussions of the concept of prolepsis, which he rendered as "pre-conception" or "pre-sensation".

In the dialogue *The Moralists: A Philosophical Rhapsody*, Theocles (who represents Shaftesbury) acknowledges that beasts have instincts that man

[17] Although Shaftesbury's *Characteristicks of Men, Manners, Opinions, Times* was published in 1711, his *An Inquiry Concerning Virtue or Merit* was first published in 1699 and his *The Moralists: A Philosophical Rhapsody* in 1703–1704. See Rivers (2000, 14–15, 101). Leibniz read the 1711 publication and his remarks on it were published by Pierre Coste in 1715; see Leibniz (1969, 629–635).

[18] Herder mentions Shaftesbury by name in his 1765 treatise *Wie die Philosophie zum Besten des Volks allgemeiner und nützlicher werden kann* (see Herder 1985, 108) and, in addition to mentioning him several times, provides excerpts in English from Shaftesbury in the first collection of his *Fragmente über die neuere deutsche Literatur* (1767; see Herder 1985, 321).

does not, such as "*Pre-sensations*", according to which, for example, pregnant females have "a clear Prospect or *Pre-sensation* of their State which is to follow; know what to provide, and how, in what manner, and at what time" (Shaftesbury 2001, vol. II, 173). But Theocles immediately considers why the same cannot be attributed to humankind and muses about what use pre-sensations could have for a species that has "what is better, in another kind [i.e.] Reason and Discourse [to] instruct them" (Shaftesbury 2001, vol. II, 174). Theocles, however, proceeds to detail just such a pre-sensation that is activated by the helplessness and weakness for which the human infant is unique among the young of all species: "[d]oes not this *Defect* engage him the more strongly to Society, and force him to own that he is purposely, and not by accident, made rational and *sociable*[?] Is not both conjugal Affection, and natural Affection to Parents, Duty to Magistrates, Love of a common City, Community, or Country [...] deduc'd from hence, and founded in these very *Wants?*" (Shaftesbury 2001, vol. II, 174). Shaftesbury is making a close connection between the natural indigence of the human infant and the activation of the "natural affections" that he attributes to human beings (and animals) that operate pre-reflectively and naturally incline human beings to virtuous and good actions that serve the preservation of the species (Shaftesbury 2001, vol. II, passim). Later in the dialogue, he returns to the notions of "anticipating *Fancys, Pre-conceptions,* or *Pre-sensations*", with Theocles announcing to an enthusiastic Philocles "I [...] will endeavour to show you that the same *Pre-conceptions*, of a higher degree, have place in human Kind" (Shaftesbury 2001, vol. II, 230). Theocles proceeds to explain how human beings, uninstructed, are naturally inclined to find beauty in figures as well as in actions:

> No sooner the Eye opens upon *Figures*, the Ear to *Sounds*, than straight *the Beautiful* results, and *Grace* and *Harmony* are known and acknowledg'd. No sooner are ACTIONS view'd, no sooner the *human Affections* and *Passions* discern'd (and they are most of 'em as soon discern'd as felt) than straight *an inward* EYE distinguishes, and sees *the Fair* and *Shapely, the Amiable* and *Admirable*, apart from *the Deform'd, the Foul, the Odious,* or *the Despicable.* How is it possible therefore not to own, "That as these Distinctions have

> their Foundation *in Nature*, the Discernment it-self is *natural*, and from NATURE *alone*?" (Shaftesbury 2001, vol. II, 231)

The natural capacities identified here are, of course, the sense for beauty and the moral sense that is closely related to it and through which beautiful actions are interpreted to be good, noble, and honest, these senses in turn having their origin in the pre-conceptions with which human beings are endowed by God/Nature and that are fundamentally related to the natural affections. Now Shaftesbury, like Leibniz, was sensitive to the critique of Locke (who had in fact been Shaftesbury's mentor) that this was to slide into the doctrine of innate ideas. In a key passage from his *Miscellaneous Reflections* in the third volume of *Characteristicks*, Shaftesbury quotes at length the words of his and Locke's mutual friend Jean Le Clerc "concerning the *natural Ideas* and the *Pre-conceptions* or *Pre-sensations* of this kind; the *Προλήψεις* [prolepseis]":

> For although, if we should speak accurately, there may be no conceptions imprinted on our minds by nature, nevertheless no one would deny that the faculties of our minds have been shaped by nature so that as soon as we start to use reason we begin to distinguish in some fashion truth from falsity, evil from good. The appearance of truth is always pleasing to us; on the other hand that of mendacity is displeasing and certainly we prefer honour to disgrace on account of the seeds planted in us which eventually spring up into the light at a time when we are able to reason; and then when the richer fruits mature by which we reason better, we are guided for public duty and education. (Shaftesbury 2001, vol. III, 130, 131)[19]

The allusion to Seneca's "seeds of knowledge" would have pleased Shaftesbury, who studied the Stoics intensively, extensively quoted from them in his notebooks, and modelled his own philosophy on them (Rivers 2000, ch. 2, passim).

[19] *Quamvis enim nullae sint, si adcuratè loquamur, notiones à natura animis nostris infixae; attamen nemo negárit ita esse facultates Animorum nostrorum naturâ adfectas, ut quàm primùm ratione uti incipimus,* Verum *à* Falso, Malum *à* Bono *aliquo modo distinguere incipiamus. Species* Veritatis *nobis semper placet; displicet contra* Mendacii: *Imo &* HONESTUM INHONESTO *praeferimus; ob* Semina *nobis indita, quae tum demum in lucem prodeunt, cum ratiocinari possumus, eoque uberiores fructus proferunt, quo melius ratiocinamur, adcuratioreque institutione adjuvamur.*

Indeed, it was not Shaftesbury's appeal to the Stoic notion of prolepsis and corresponding dispositional innatism alone that was the source of his appeal for Herder; centrally important was also how Shaftesbury incorporated this into a rich philosophy of nature. In both the *Inquiry* and *The Moralists*, Shaftesbury explains how the "natural affections" in both animals and human beings are fitted both to the good of their individual "system" qua organisms and species, and to the "SYSTEM *of all Things, and a Universal Nature*" (Shaftesbury 2001, vol II, 9, 12). *The Moralists* is an extended paean to the providential natural order of the universe that had a profound effect on the young Herder, who was even moved to put a section of the dialogue to verse, which he subsequently published as "Shaftesburys Naturhymnus" in the 1790 edition of his treatise on Spinoza, *Gott: einige Gespräche* (Herder 1877–1913, vol. 27, 418), and in which he defends his dynamic conception of organic forces. This corresponds to a further dimension of Shaftesbury's Stoicism, it being well-established that Stoics believed nature to be providential. In the words of Dominic Scott, "for the Stoics Nature and God are equivalent" (Scott 1988, 147).[20] This brings us back to the question of psychological versus epistemological innatism. For the Stoics, it is not only a fact that human beings are innately disposed to develop certain moral notions, it is further the case that these notions can be taken to be trustworthy in virtue of the fact that since human beings themselves are parts of nature, their innate tendencies are equally natural and more likely than not will promote rather than detract from nature as a whole (Scott 1988, 147). It is precisely this idea that we find reflected and repeated in Shaftesbury. And his modern interpretation and elaboration of these key Stoic ideas we see, in turn, reflected in and indeed pervading Herder's philosophy as whole, from his reflections on moral philosophy, to his defence of natural sociability against Rousseau in the treatise on language, to his detailed elaboration of the history of the providential order as a natural whole and his theory of organic forces in the *Ideen zur Philosophie der Geschichte der Menschheit* and *Gott: einige Gespräche*.[21]

---

[20] This is reminiscent, of course, of Herder's famous cautionary words from the Preface to the *Ideen zur Philosophie der Geschichte der Menschheit*: "Niemand irre sich daher auch daran, daß ich zuweilen den Namen der Natur personifiziert gebrauche. Die Natur ist kein selbstständiges Wesen; sondern Gott ist Alles in seinen Werken" (Herder 1989, 17).

[21] For discussion of these three themes, respectively, see DeSouza (2012a, 2014, 2018).

3 Baumgarten

Thus far, I have sought to show how in the period of his earliest philosophical activity, from the mid to the late 1760s, Herder was exposed to and assimilated the modern forms of Stoic dispositional innatism that he encountered in Leibniz and Shaftesbury. It was noted that Herder integrated this dispositional innatism into Leibniz's theory of types and levels of knowledge, associating it, in particular, with the obscure or *dunkel*. The latter was commonly associated with the lower, sensuous regions of the soul, in contrast to the upper, rational regions that were the realm of clear and distinct knowledge. Now in this same period that we have been considering, Herder became enamoured of the philosophy of Alexander Gottlieb Baumgarten precisely for the promising reason that Baumgarten had made his intention to study of the lower regions of the soul explicit. In the closing paragraphs of the *Meditationes philosophicae de nonnullis ad poema pertinentibus* from 1735, which was devoted to poetry, Baumgarten broadened his perspective to the lower faculty of knowledge, of which poetry, he claimed, is a product. After acknowledging that the upper faculty of knowledge is equipped with the instrumental tool of the science of logic to direct it in the discovery of truth, Baumgarten asserts that psychology provides firm principles for asserting the rightful claim to existence of a science directing the lower faculty of knowledge, the science of how something is sensuously known. This science, whose objects are *aistheta*, is none other than aesthetics (Baumgarten 1983, §§115–116). Instrumental philosophy is henceforth no longer to be synonymous with logic, for it must now also make room for aesthetics. In so doing, Baumgarten was responding not only to the complaints, like those of Johann George Sulzer, about the neglect of the lower, sensuous regions of the soul, or to the call of Georg Bernhard Bilfinger for a philosophical theory of sensuousness analogous to what Aristotle's logic provided for the intellect. He was also heeding the demand of Wolff himself for a philosophy of the liberal arts, which would consist of a science of the general rules of grammar, rhetoric, poetry, and of other such arts (Wolff 1996, §72). What Herder hoped to find, then, in Baumgarten's *Aesthetica* (1750, 1758), the work in which Baumgarten would present his aesthetic theory, was what he had failed to find in Wolff, namely, an investigation of the lower faculties of the soul that would in turn prove genuinely useful and in which Baumgarten would turn to the "obscure perceptions" of the "FUNDUS ANIMAE", in the words of his *Metaphysica*.[22] Indeed, as Herder writes, full of expectation, in his reflections on the *Meditationes*: "[i]m Geiste des Menschen, das war Baumgartens große Ahndung, in der Seele muß der Poesie ein Gebiet des Eigentums zuerkannt, und genau angewiesen werden können. Hier müssen Kräfte liegen, die dieselbe hervorbrachten, und Kräfte, die sie wieder beschäftiget. Hin also in diese dunkle Gegenden, um aus ihnen, wie aus einer Wunderhöhle Nachrichten zu bringen, wo diese Göttin wohne" (Herder 1985, 684). This enthusiasm, however, soon turned to disappointment, and then detailed critique, articulated in writings from 1767, as Herder came to realize that what he had hoped to uncover in Baumgarten's study of aesthetics was far from a study of the lower regions of the soul that genuinely sought to come to grips with them. "Eine Aesthetik des Autors aber: sie wäre so wahr, als sie wollte, ist im höchsten Grade unphilosophisch: da sie alle philosophischen Begriffe der Aesthetik, die sie entwickeln soll, der Metaphysik voraussetzt, sie ihr nachbetet, und eine Menge leere, falscher, zu feiner Folgerungen herauszieht" (Herder 1985, 673). Now I have reconstructed the details of Herder's careful critique elsewhere.[23] What I wish to do in the final section of this paper is present a brief outline of the *alternative* conception of a genuine aesthetics that Herder's presents in response, in light of our discussion so far.

Baumgarten's error, in Herder's eyes, was to begin his analysis from the products of beauty and from beautiful thinking, on the basis of which reified and universal rules are abstracted. This forms the wrong starting point, and this error, in turn, is premised on a problematic model of the soul. In a word, Herder rejects Baumgarten's definition of aesthetics as the science of sensuous knowledge with which the *Aesthetica* begins (Baumgarten 2007, §1, 10–11). "Die Schönheit ist also nicht ein Stück der sinnlichen Erkenntnis", Herder protests (Herder 1985, 673). First, to begin from the idea that beauty is a form of *knowledge* is to skew our

---

[22] Baumgarten, *Metaphysica*, §511.

[23] Cf. DeSouza (2016a), which I draw on in what follows.

understanding of beauty, from the outset, towards clear and distinct knowledge. For as Baumgarten makes clear in the *Aesthetica*, aesthetics as the science of sensuous knowledge begins from a conception of beauty as the perfection of confused but *clear* knowledge. Perfection is the agreement of several things according to the same ground, and when this perfection is sensuous, its sensuous and clear perception is experienced as beauty. For Herder, Baumgarten's conception of aesthetics as the science of sensuous *knowledge* is inherently problematic because it begins from the idea of beauty as a form of clear knowledge, that is, perceptions that are apperceived, recollected, and that may be recognized again. But for aesthetics to start from an investigation of beauty so defined is for aesthetics to begin from what Herder calls a "Mittelding"—something halfway between the obscurest perceptions, on the one hand, and the clearest and most distinct, on the other (Herder 1985, 671). On Herder's conception, aesthetics must, as a study of the sensuous, lower region of the soul or *Seelenkräfte*, begin not from where knowledge begins, but far lower, with the most obscure regions. For Herder, aesthetics, as a study of the sensuous, lower regions, begins not from clear knowledge then, but from feeling, *Gefühl* (Herder 1985, 672).

Baumgarten makes a distinction between natural and artificial aesthetics, Herder's critical discussion of which takes us deeper into his alternative proposal for aesthetics. Every individual has an innate ability to think beautifully (innate aesthetics) and they can also naturally learn rules for doing so through the mere use of this faculty (acquired aesthetics)—this is what Baumgarten calls "natural aesthetics" and it is analogous to the natural logic that each person has through which they are able to represent the nexus of the world to themselves distinctly (Baumgarten 2007, §2, 11–13). The acquired part of natural aesthetics is then further subdivided into "eine lehrende" and "eine ausübende" Aesthetik. The former is one's confused knowledge of the rules of thinking beautifully, while the latter is the ability to apply these rules in practice. But just as one can study the science of logic and explicitly learn its rules and their foundation and thus improve one's ability to think distinctly (what Wolff calls "artificial logic"), so too can one study the science of aesthetics and develop a *distinct* knowledge of its rules and their foundation and improve one's ability to think beautifully—and this is what Baumgarten calls "artificial aesthetics" (Baumgarten 2007, §3, 12–13). The latter is based on a study of the beauty of knowledge in the liberal arts in general and it takes up most of the 900 paragraphs of the *Aesthetica*. Aesthetics is thus based on a natural possession of the art of thinking beautifully that is in turn both further directed and developed as well as grounded scientifically by artificial aesthetics. Now Herder, compelled to use these terms of art by Baumgarten, by contrast has a much deeper and broader conception of natural aesthetics, which differs from artificial aesthetics not by mere degrees, as with Baumgarten, but essentially. "[Natürliche Aesthetik] *immer* habitus, [künstliche Aesthetik] aber scientia ist; jene in *Empfindungen* und *dunklen* Begriffen *wirkt*; diese in *Sätzen* und deutlichen *Begriffen lehrt, überzeugt*" (Herder 1985, 660). The innate aesthetics of a pre-reflexive capacity for thinking beautifully in Baumgarten is, in Herder, transformed: "Alle Menschen haben aestheticam *connatam* [i.e., innate aesthetics], da sie *alle* als *sinnliche* Tiere *geboren werden*, und weil sie mehr Tiere als Geister sind" (Herder 1985, 660). Thinking—that is, knowledge—is thus not first in humans, and beautiful knowledge is thus not first in aesthetics. Rather, aesthetics as a study of the sensuous lower *Seelenkräfte* starts from the human/animal's obscurest *Gefühl*, which, as it becomes more lively, begins to approach the "dunklen *Empfindungen des Selbst*" that constitutes the lowest level of apperception—the starting point, below this, then, is shrouded in total obscurity (Herder 1985, 676). The examination of these sensuous regions of the soul, Herder declares, "ist ein Stück von der *nötigsten* Anthropologie, da in dem *Grunde* der Seele unsere *Stärke* als *Menschen* besteht" (Herder 1985, 665). Herder thus resolutely rejects Baumgarten's assertion in the *Aesthetica* that sensuous knowledge that "is so hidden, that it either remains fully obscure to us or that we can only know through an intuitive intellectual grasping [non nisi intelligendo possimus intueri] is not the concern of the aesthetician as such" (Baumgarten 2007, §15, 21). Herder is, very much to the contrary, advocating a plunging headlong into this region of the soul out of which emerge dispositions, inclinations, drives, and so on implied by the word "Stärke". He elaborates:

> Da im verborgensten Grunde der Seele die mächstigsten Triebfedern liegen, von denen die bekannteren getrieben werden: so wäre es eine vergebne

> Arbeit, von diesem Mitteldinge [i.e. beauty] an; an beide Enden zu kommen; hier grabe also der Aesthetiker [...] Vielleicht ists völlig non-sens, wenn der Autor das aus der Sphäre des Aesthetikers bannt, quae non nisi intelligendo possimus intueri, bloß das ist vor ihn quae non nisi intelligendo possimus intueri. (Herder 1985, 671)

Herder takes the study of this neglected region of the soul so seriously that he even proposes assigning it its own logic. The standard conception of logic, which he associates with Baumgarten and which is a part of metaphysics, is characterized as neither organic nor possible *a priori* since it limits itself to effects, to "die *Wirkung*" (Herder 1985, 667). This logic reifies or hypostasizes the *results* "[des] *denkenden* Selbst[s]" into timeless and universal rules. Herder contrasts this with a second conception of logic, which he construes as a part of psychology and which *is* organic in that it is a genetic examination starting from the *Seelenkräfte* and yielding, in fact, multiple logics: "es ist also vor jede Kraft eine *Logik* (nicht aber als Besserungs-, sondern *Untersuchungs*logik) möglich; die aber je tiefer es zu den Seelenkräften kommt, desto weniger in Regeln wird ausschweifen können" (Herder 1985, 667). The object of this kind of logic is to examine how the various forces and powers of the soul operate and develop, from the bottom up. Undergirding it all would be a kind of foundational logic: "[d]ie Erste Logik untersuche den *Menschen*, das Tier, <das> vernunftähnlich nach Gedanken und Trieben handelt; und ist die kürzeste, die schwerste, die wichtigste, der Grund der andern, von der man noch nicht den Schatten hat, da doch das *Wahre* der andern sich auf sie bezieht" (Herder 1985, 667). This is a logic that is descriptive, not normative, that is oriented towards a primordial knowing-how rather than an evolved knowing-that, and whose object is the development of the human soul and its capacities from their origins. It stands in contrast to the traditional conception of logic which observes "den *Menschen* mit den *Auswüchsen* seiner *Erkenntnis* und Begierden; als ein vernünftelndes und feines Tier" (Herder 1985, 667). The aesthetics Herder favours "steige[re] [...] vom heißesten Affekt, der ganzesten Empfindung immer mehr zum Ideenartigen" and it is only where feeling almost reaches consciousness that "die eigentliche Logik" begins (Herder 1985, 667–668). In sum, such a logic fully fleshed out, then, would start from organic forces in a living being that express themselves at each level of organization in different ways; to name only a few of these levels: (i) attraction and repulsion, expansion and contraction, muscular irritability and nervous sensation; (ii) upwards to obscure feelings, inclinations, drives, instincts, moral feelings/sense, sensuous pleasure, the perception of beauty, and (iii) on finally to the various stages of clear and distinct knowledge. And while it is within the domain of (ii) that Stoic dispositional innatism most obviously exercises its influence, as we saw above in Sect. 1 on Leibniz, it also extends to (iii) and to fundamental theoretical concepts that underpin initially pre-reflectively human behaviour and knowledge.

With Herder's detailed proposals for a more genuine and adequate aesthetics, elicited from his study of Baumgarten's writings, we see the rudiments of his theory of organic forces that will come to full fruition in the first two parts of the *Ideen zur Philosophie der Geschichte der Menschheit* (1784). In moving from a substance-based ontology to his particular force-based ontology, Herder was both contributing to and reflecting the spirit of his times in which vitalism and the study of living nature were taking centre stage in European thought. We can now also see more clearly why the concept of drive is not ontologically basic for Herder. In a word, drives *qua* innate dispositions emerge out of forces. For it is only at a certain stage of organic development that begins with the unfolding of soul-forces that drives arise and become operational.

There were, of course, a great number of intellectual influences and sources behind Herder's philosophy of life and theory of organic forces. My objective in this paper has been to begin to trace the role played by just one of those influences in Herder's thought: the tradition of Stoic thought, in particular, of Stoic dispositional innatism, especially as transmitted by Leibniz and Shaftesbury.

References

Adler, Hans. 1990. *Die Prägnanz des Dunklen: Gnoseologie, Ästhetik, Geschichtsphilosophie bei J. G. Herder*. Hamburg: Felix Meiner Verlag.

Baumgarten, Alexander Gottlieb. 1983. *Meditationes philosophicae de nonnullis ad poema pertinentibus. Philosophische Betrachtungen über einige Bedingungen des Gedichtes*. Edited by Heinz Paetzold. Hamburg: Meiner.

———. 2007. *Aesthetica. Ästhetik*. Vol. I. Edited by Dagmar Mirbach. Hamburg: Meiner.
Bui, Hien. 2017. A Case for Dispositional Innatism. *Res cogitans* 8 (1): 81–89.
DeSouza, Nigel. 2012a. Language, Reason, and Sociability: Herder's Critique of Rousseau. *Intellectual History Review* 22 (2): 221–240.
———. 2012b. Leibniz in the Eighteenth Century: Herder's Critical Reflections on the *Principles of Nature and Grace*. *British Journal for the History of Philosophy* 20 (4): 773–795.
———. 2014. The soul-body Relationship and the Foundations of Morality: Herder Contra Mendelssohn. *Herder Yearbook* 21: 145–161.
———. 2016a. Aukflärerische Selbstentwürfe. In *Herder Handbuch*, ed. Stefan Greif, Marion Heinz, and Heinrich Clairmont, 71–121. Paderborn: Wilhelm Fink Verlag.
———. 2016b. Deux périodes et métaphysiques de l'Aufklärung: Herder et sa critique de Wolff. In *Les métaphysiques des Lumières*, ed. Pierre Girard, Christian Leduc, and Mitia Rioux-Beaulne, 191–205. Paris: Classiques Garnier.
———. 2017. The Metaphysical and Epistemological Foundations of Herder's Philosophical Anthropology. In *Herder: Philosophy and Anthropology*, ed. Anik Waldow and Nigel DeSouza, 52–71. Oxford: Oxford University Press.
———. 2018. Herder's Theory of Organic Forces and its Kantian Origins. In *Kant and His German Contemporaries, Volume II: Aesthetics, History, Politics, and Religion*, ed. Daniel O. Dahlstrom, 109–127. Cambridge: Cambridge University Press.
Dreike, Beate. 1973. *Herders Naturauffassung in ihrer Beeinflussung durch Leibniz' Philosophie*. Wiesbaden: Franz Steiner.
Dyson, Henry. 2009. *Prolepsis and Ennoia in the Early Stoa*. Berlin: De Gruyter.
Haym, Rudolf. 1885. *Herder nach seinem Leben und seinen Werken dargestellt*. Zweiter Band. Berlin: R. Gaertners Verlagsbuchhandlung.
Heinz, Marion. 1994. *Sensualistischer Idealismus: Untersuchungen zur Erkenntnistheorie und Metaphysik des jungen Herder (1763–1778)*. Hamburg: Felix Meiner Verlag.
———. 2020. Sympathie. Zur Palingenesie stoischen Denkens im Herders *Kalligone*. In *Herder on Empathy and Sympathy: Einfühlung und Sympathie im Denken Herders*, ed. Eva Piirimäe, Liina Lukas, and Johannes Schmidt, 50–73. Brill: Leiden.
Herder, Johann Gottfried. 1846. *Johann Gottfried Herders Lebensbild. Sein chronologisch-geordneter Briefwechsel*. Vol. II. Edited by Emil Gottfried von Herder. Erlangen.
———. 1877–1913. *Herders sämtliche Werke*. Berlin: Weidmannsche Buchhandlung.

———. 1979. *Der handschriftliche Nachlass Johann Gottfried Herders: Katalog*. Edited by Hans Dietrich Irmscher and Emil Adler. Wiesbaden: Otto Harrassowitz Verlag.

———. 1984. *Briefe*. Vol. I. Edited by Wilhelm Dobbek and Günter Arnold. Weimar: Böhlhaus Nachfolger.

———. 1985. *Frühe Schriften 1764–1772*. Edited by Ulrich Gaier. Frankfurt am Main: Deutscher Klassiker Verlag.

———. 1987. *Herder und die Anthropologie der Aufklärung*. Edited by Wolfgang Pross. München: Carl Hanser Verlag.

———. 1989. *Ideen zur Philosophie der Geschichte der Menschheit*. Edited by Martin Bollacher. Frankfurt am Main: Deutscher Klassiker Verlag.

———. 1994. *Plato sagte: daß unser Lernen bloß Erinnerung sei*. In *Sensualistischer Idealismus: Untersuchungen zur Erkenntnistheorie und Metaphysik des jungen Herder (1763–1778)*. Hamburg: Felix Meiner Verlag, 175–182.

Klein, Jacob. 2011. Review of: Henry Dyson, *Prolepsis and Ennoia in the Early Stoa*. *Journal of the History of Philosophy* 49 (I): 115–116.

Leibniz, G. W. 1765. *Oeuvres philosophiques latines et françoises de feu M. de Leibnitz*. Edited by R. E. Raspe. Amsterdam: Jean Schreuder.

Leibniz, G. W. 1969. *Philosophical Papers and Letters*. Edited by Leroy E. Loemker. Dordrecht: D. Reidel Publishing Company.

Pénisson, Pierre. 2002. Trieb et énergie chez Herder. *Revue germanique internationale* 18: 45–52.

Piirimäe, Eva. 2020. Introduction: Historicism, Stoicism and Contemporary Perspectives. In *Herder on Empathy and Sympathy: Einfühlung und Sympathie im Denken Herders*, ed. Eva Piirimäe, Liina Lukas, and Johannes Schmidt, 1–36. Brill: Leiden.

Rivers, Isabel. 2000. *Reason, Grace, and Sentiment: A Study of the Language of Religion and Ethics in England 1660–1780, vol. II, Shaftesbury to Hume*. Cambridge: Cambridge University Press.

Scott, Dominic. 1988. Innatism and the Stoa. *The Cambridge Classical Journal* 34: 123–153.

Shaftesbury, Anthony Ashley Cooper, Third Earl of. 2001. *Characteristicks of Men, Manners, Opinions, Times*. Edited by Douglas Den Uyl. Indianapolis: Liberty Fund.

Wolff, Christian. 1996. *Discursus praeliminaris de philosophia in genere. Einleitende Abhandlung über Philosophie im Allgemeinen*. Edited by Günter Gawlick and Lothar Kreimendahl. Stuttgart: frommann-holzboog.

Zammito, John. 2017. Herder between Reimarus and Tetens: The Problem of an Animal-Human Boundary. In *Herder: Philosophy and Anthropology*, ed. Anik Waldow and Nigel DeSouza, 127–146. Oxford: Oxford University Press.

5

The Economy of the *Bildungstrieb* in Goethe's Comparative Anatomy

Andrew Cooper

1 Introduction

Goethe's role in the formation of evolutionary science is a contentious matter. In his writings on botany and zoology, Goethe aimed to replace the established practice of natural history, which identified class boundaries according to manifest characteristics, with the search for a type (*Typus*) or building plan (*Bauplan*) that lies behind the development of living things. In the mid-nineteenth century, William Whewell (1847, III 490) claimed that Goethe's shift from isolated characteristics to morphological type was "acknowledged by the best authorities" during the early development of biology. Charles Darwin (1959, 61) confirmed Whewell's observation a few years later by naming Goethe as "an extreme partisan" of his transmutational view of species. Hermann von Helmholtz (1865, 31–55) claimed that Goethe's morphology made it possible for Darwin's evolutionary theory to be taken up so quickly in Germany. In

A. Cooper (✉)
University of Warwick, Coventry, UK
e-mail: andrew.j.cooper@warwick.ac.uk

M. Kisner, J. Noller (eds.), *The Concept of Drive in Classical German Philosophy*,
https://doi.org/10.1007/978-3-030-84160-7_5

the twentieth century, however, historians of biology tended to reject such praise as misguided. Ernst Mayr (2000, 327) contrasts Goethe's morphology with the population dynamics of Darwinian physiology, presenting Goethe as an essentialist who believed in the unfolding of immanent potentialities. George Wells (1978, 45–46) rejects any serious contribution Goethe might have made to biology on the grounds that he "was unable to accept the possibility of large scale evolution".

In this chapter I reconsider the early reception of Goethe by Whewell, Darwin and Helmholtz by returning to an overlooked concept in his morphological writings: the "economy of nature [*Ökonomie der Natur*]" (LA I 9:204).[1] During the late 1780s, Goethe became critical of the sciences concerned with the parts and process of living bodies, which examined organic forces by analogy with Newtonian gravity. In the early 1790s, he began to experiment with a new analogy drawn from Carl Linnaeus' *Oeconomia Naturae* (1749), in which the populations of various species are considered within a mutually beneficial and self-regulating economy.[2] The relations between these populations are guided by a distributive principle Goethe termed the "formative drive [*Bildungstrieb*]". On the level of the singular organism, the comparative anatomist does not consider the relation between the parts is as a function of an external force or preestablished end. The formative drive freely distributes available resources according to an idea, which, like an artist with a finite quantity of materials, can nevertheless produce a potentially infinite variety of shapes. On the macro level of species populations, the comparative anatomist can consider nature as an interlaced network of household budgets, wherein the various structural plans are codetermined by the interaction between the formative drive and external conditions.

My aim in what follows is to show that Goethe's economic analogy replaces the Newtonian model of force with an experimental conception of the formative drive, thereby opening a large-scale morphological

[1] Citations to Goethe will be in-text, and use the following abbreviations: LA = Leopoldina Ausgabe (Goethe 1947); WA = Weimar Ausgabe (Goethe 1888); ZM = *Zur Morphologie* (Goethe 1817).

[2] "By the Economy of Nature we understand the all-wise disposition of the creator in relation to natural things, by which they are fitted to produce general ends, and reciprocal uses" (Linnaeus and Biberg 1749, §1, c.f. §20).

programme of research. This feature of his work was rightly picked up by Whewell, Darwin and Helmholtz, and yet was overlooked by later biologists given the emphasis on population genetics following the Modern Synthesis. I begin in Sect. 2 by contextualising Goethe's programme for comparative anatomy within his study of metamorphosis. In Sect. 3 I identify his search for a typological norm in the late 1780s, which begins during his first Italian journey (1786–1788) and culminates in *Essay to Explain the Metamorphosis of Plants* (1790). In Sect. 4 I then discern a shift that begins with "First Outline of a General Introduction to Comparative Anatomy" (1795), in which Goethe introduces the formative drive to integrate his earlier work on typology with morphological considerations. In the final section I consider his presentation of the development of morphology in his brief essay "Formative Drive" (1810). While Goethe does not propose a mechanics of transmutation, I contend that his economic analogy marks an important advance in the study of organic form, for it places comparative anatomy within a biodynamic system of exchange.

2 From Classification to Typology

When Goethe finally published his writings on morphology in 1817, he included a prefatory essay that names Linnaeus as the key to his intellectual formation (ZM I xxii). The essay invites us to consider how Linnaeus' attempt to bring the inchoate study of natural history into a single programme of research shaped Goethe's thinking long before his writings on comparative anatomy. During the early 1780s, Goethe undertook a detailed study of *Systema Naturae* (1735), a text that guides the natural historian beyond surface characteristics to intrinsic qualities manifest in reproduction. His ongoing interest in Linnaeus' dynamic view of the natural system is reflected in the fact that, when he slipped away from Weimar in September 1786 to embark on his first Italian journey, Linnaeus' *Fundamenta botanica* was one of the few books he took with him (Kuhn 1987, 5). Throughout his travels, Goethe frequented the botanical gardens of Padua, Bologna and Palermo, where he developed his skills in sketching individual plant specimens. Yet his detailed sketches led him to doubt the clean boundaries of the Linnaean system:

> When I discovered on the stem of the very same plant leaves that were first round, then indented, and finally almost feathered, which later contracted, simplified, and turned into scales, and at last disappeared altogether, I lost the courage to drive in a stake, or to draw a dividing line. It seemed to me that the problem of designating the genera with certainty, and of arranging species under them, was unsolvable. (WA II 6:117)

In place of the isolated elements chosen by Linnaeus, Goethe began to consider the plant as a living whole, extended in time and governed by a movement from whole to parts. Where Linnaeus searched for increasing definition, Goethe's sought to unite and reduce variation to a single type, in relation to which all existing organisms stand as tokens. While none of the tokens give full expression to the type, the type nonetheless exists in the token, allowing the keen observer to identify specimens as belonging to the same genera. A species is neither a logical category nor a physical line of decent, but an empirical actuality *within* each instance.

Goethe's interest in the products of nature corresponded with a growing frustration with his own productivity as an artist. In contrast to artworks, which rarely gave full expression to the artist's idea, natural products reach perfection in every instance. Goethe reasoned that if human art arises from a principle, then works of nature could arise in an analogous way. In a letter written from Naples in June 1787, he announced to Charlotte von Stein that he was very close to discovering the principle of plant growth, which he termed the *Urpflanze*:

> Tell the Herders I am very close to the secret of plant organization. […] The *Urpflanze* will be the most extraordinary creation in the world, one that nature herself might envy me. With this model [*Muster*] and the key to its use, one could go on forever inventing plants, which means that, even if they did not exist, still could exist not just as painted or poetic shadows and illusions, but rather as having an inner truth and necessity. (LA II 9A:365)

Goethe does not present the *Urpflanze* as an existing plant or an ancestral form but as something abstracted from empirical particulars. His claim that nature herself might envy him indicates that the *Urpflanze* is not a natural product but a general plan that is never fully realised in any specific form (Brady 1987, 269). In contrast to Linnaeus' static conception of shape, Goethe's model aims to capture the transformative process that leads both to and from the product, a movement that is distinct from the causal interaction between material particles. The *Urpflanze* acts as an interface between the poetic and the scientific, capturing the inner truth and necessity of a natural product that eludes the scientist concerned only with contact forces (Holland 2009, 20). If the scientist is not simply to discern the metamorphoses of its products, as Linnaeus had done, but also the thing that changes throughout these metamorphoses, she must adopt the standpoint of the artist. Once she has grasped the inner truth of physiological movement, she can produce an infinite variety of unseen forms in her mind, not as fiction but as potential reality.

Goethe's search for the *Urpflanze* consolidated during a visit to the gardens of Padua in September 1786 (WA I 30:89). As he examined a towering palm in the greenhouse, he was able to see the various stages of leaf development at a single moment (ZM I xxviii). The attempt to grasp all stages together led him to postulate a fundamental organ as the force governing the generation of the various parts of the plant. This organ was not a projected idea but rather an actual part that plays out in all the parts of the plant, namely, the leaf. In his notebook he sketched a single form taken by the various alterations between developmental stages, proclaiming that "all is leaf, and through this simplicity the greatest multiplicity is possible" (LA II 9A:58). Goethe refined his account of the leaf during a visit to the gardens of Palermo later that year. He describes the leaf as a force that governs the generation of the various parts of the plant: "Forwards and backwards the plant is always only leaf" (WA I 31:239f.). In contrast to the accepted evolutionist theories, in which a physical germ expands due to external forces, the leaf is a principle reproduced through the generative process, expanding and retracting as the plant moves through various stages of development.

Throughout his travel writings, Goethe rehearsed several ways of presenting the *Urpflanze* and its organ, the leaf. At some points he describes the *Urpflanze* as a model or type (*Muster, Typus*) that can be intuited by carefully observing the growth and development of flowering plants and ferns. Elsewhere he describes it as a formula, rule or law (*Formel, Regel, Gesetz*) that governs the formation of plants, discovered by comparing the kinds of growth between species. Each formulation signals an attempt to combine external form with inner process. Yet as Goethe's view developed, the ontological status of the *Urpflanze* remained unclear (Larson 1967, 595). Throughout his notebooks there is a conspicuous lack of reflection on scientific method and the causality of ideas (see Boyle 2000, II 597). Robert Richards (2002, 422) presents the young Goethe as a Spinozistic realist, for he considered the laws governing plant development to be expressed in the archetype. Like Spinoza's "adequate idea", the *Urpflanze* embodies the essence of natural objects, and is discovered through observation and research. Yet in contrast to Spinoza, Goethe's *Urpflanze* was not a static force but an active power that gives rise to an endless variety of forms.

3 From Typology to Comparative Anatomy

Goethe's return to Weimar for the summer of 1788 was marked by severe disappointment. In contrast to the rich natural vista of the south, his homeland seemed "formless" and "dark". Nevertheless, in the following years Goethe made significant progress in his botanical work. In November 1789 he began writing a new essay that accounted for the physiology of plants according to a fundamental type, which he published the following Easter as *Versuch die Metamorphose der Pflanzen zu erklären* (*Essay to Explain the Metamorphosis of Plants*, 1790). In a letter written to Karl Ludwig von Knebel in June, Goethe declared that the essay marked the beginning of a new preoccupation with natural science, in which he began to harmonise the search for a primal form with the broader interest in generative processes (WA IV 9:213). His aim was to outline a new method of comparison that does not emphasise structural analogy, such as the comparison of organ with organ or shape with shape, but rather the dynamic transformation of structurally similar parts along a single axis. The result would be a vivid description of the archetypal plant, which adopts different formations over the course of time.

The explanation presented in *Metamorphosis of Plants* does not follow the Newtonian physiology advocated by Albrecht von Haller, which accounts for morphological change according to living forces that cause a performed miniature to expand into a developed form. Neither does it follow Georges Buffon's mechanical vitalism, in which the shape of a living being is the result of organic matter pressed into an inner mould by a penetrating force. Goethe's fundamental building block of vegetable life, the leaf, is not a prestructured form that works behind the morphological process, but rather a principle of change manifest in the variety of forms assumed by the different parts of the plant. The essay is an explanation to the extent that it describes the process by which the leaf comes to expression in the various parts of the plant through a temporal process of expansion and contraction, enabling the reader to discern that process in *any* plant (LA I 9:59). Its detailed description of plant growth evokes a visualisation of metamorphosis, allowing one to grasp the internal relations behind any existing instance. Metamorphosis is the "regular" or "progressive" growth of the plant from seed to leaf and to fruit as the "transformation of one shape [*Gestalt*] into another", climbing a "spiritual ladder" to the pinnacle of nature: "the propagation [*Fortpflanzung*] through two sexes" (LA I 9:24). The leaf is neither an eternal form nor a physical part, but a principle that transmutes the physical structures of the plant (Benn 1949, 176).

Following the publication of *Metamorphosis of Plants*, Goethe returned to Italy in search of the vivacity he experienced during his first sojourn in the south. Yet this venture thoroughly disappointed his expectations, and he did not stay long abroad. Nevertheless, his second journey provided an experience that he would later recall as a pivotal moment in his intellectual development. While taking his regular stroll along the lido in Venice, Goethe came across a sheep's skull. The skull was open in such a way that enabled him to connect the sequential unfolding of the vertebrae with the development of the leaf across the axis of the plant, placing "before [his] eyes the transition from inner, unformed, organic masses, through their development outwards, via the progressive refinement of the highest formation and evolution into the most excellent sensory tools" (WA I 35,

13:15). As the plant consists of transformed leaves, so the skull is composed of metamorphosed vertebrae, providing visible traces of a homology between the bone structure in humans and animals.

By identifying a structural plan (*Bauplan*) that unites vertebrates in a single group, Goethe discovered a way to transpose his morphological thinking from plants to animals. His new insight coincided with a growing concern with the foundations of scientific knowledge and the appropriate methods of investigation, prompted by the youthful Friedrich Schiller, who pushed him to consider his search for a physiological type within the framework of Kant's transcendental epistemology.[3] Goethe was already acquainted with Kant's transcendental account of nature in *Critique of Pure Reason*, and found an ally in *Metaphysical Foundations*, where Kant banishes external causes from the consideration of matter. Yet the epistemic restrictions Kant placed on ideas failed to expunge his Spinozist commitments. In Goethe's view, Kant demonstrated that organic form cannot be explained by external causes. Natural variation must be considered as a dynamic interaction between external forces, which act on living beings, and inner purposiveness, by which the organism gives shape to itself. Organic productivity is thus akin to artistic productivity, for organisms and artworks both "exist for their own sake", existing "*for* one another, but not *because of* one another" (LA I 9:92). Kant's presentation of a shared basis for scientific and aesthetic judging assisted Goethe to see that while we must consider natural products to be shaped by their environment, the internal structure of living beings cannot be explained by causes that are external to it. Morphological concepts are thus not observed but operate as ideas that guide inquiry.[4] They cannot be seen with the physical eye but must be contemplated with the inward eye.

In the years that followed his encounter with the sheep's skull on the Venetian lido, Goethe worked extensively on sketching out a new science of morphology. Six months after his second Italian journey, he began his first attempt, "Versuch über die Gestalt der Tiere" ("Essay on the Shape

[3] See Goethe's account of his friendship with Schiller in "Glückliches Ereignis" (LA I 9:79–83).

[4] See Goethe's account of the influence of Kantian philosophy on the physiology of the 1790s in "Einwirkung der neueren Philosophie" (LA I 9:90–94).

of Animals"). In this essay, Goethe announced his intention to define the general shape (*Gestalt*) that unites the animal kingdom. His method is to examine skeletal growth in mammals in search of an "osteological type [*osteologischen Typus*]" that gives rise to the manifold of shapes (LA I 10:78). While this type resists containment in any single shape, Goethe nevertheless insists that his method is "merely empirical" (LA I 10:79). It does not explain how the bones come together in such a form but describes the process shared among all vertebrates. The "cause of the similarity of the most varied forms" is the "invariable connection of the parts with each other" (LA I 10:79). Over the next five years, Goethe composed four more essays on the general shape of mammals. In "Versuch einer allgemeinen Knochenlehre" ("Essay on a General Theory of Bones", 1794), he proposed an empirical investigation of the vertebrate skeleton, offering a detailed description of the head to the backbone (LA I 10:87f.). While he recognised the provisional nature of his studies, which focused on just one element of the developmental process, Goethe aimed to carve out an experimental path towards a *Metamorphosis of Animals* (Boyle 2000, II 256). Such a standard could not be derived from any one specimen or species, but rather as a scheme by which the skeletons of particular existing species could be compared and their variations described.

As he developed a new programme for comparative anatomy, Goethe worked on a methodological essay entitled "Versuch einer allgemeinen Vergleichungslehre" ("Essay on a General Theory of Comparison", 1790–1794). In this text we can discern the growing impact of Schiller and Kant's third *Critique* on his thinking. Goethe begins by decrying the traditional conception of living beings as products of an external purpose, which has hitherto restricted all progress in physiology when compared to the other sciences (LA I 9:118–119). When physiological movement is examined though a Newtonian analogy, the physiologist searches for laws that govern the present arrangement of the parts within an organic system. Such an inquiry remains impervious to the manner in which the parts dynamically constitute the system itself, and maintain themselves in a self-regulating structure. To examine the formative process as the result of an *inner* process that follows a single pattern, the anatomist must consider the whole as the product of the relation between its parts. Thus conceived, the primary question of comparison is how a surrounding element, such as water or air, affects the expression of that pattern. Goethe's aim is to shift the reader's perspective from a state of wonder at the fit between the functions of an organic body with its environment to an examination of that fit as a natural achievement. The inner principle determines how the general pattern is expressed in response to outer forces. An organism is thus shaped according to a dynamic interplay of internally defined needs and externally defined conditions. And Goethe does not stop with the singular organic body. The concept of a self-producing whole opens a higher level from which to consider "the organised world itself as an interrelationship of many elements" (LA I 10:122). When the world is considered as a self-producing whole, one can see that the plant kingdom is necessary for the existence of insects, oceans and rivers are necessary for fish, such that one views the "whole world of animals as a great element in which one species is created, or at least sustained, by and through another" (LA I 10:122). The conception of natural order as a dynamic, self-producing whole destroys any reference to the connections and relationships in nature as the product of an external purpose.

4 The First Outline

Goethe's "Essay on a General Theory of Comparison" is a transitionary text that stands between his earlier work on the osteological type and a new text he completed in 1795, "Erster Entwurf einer allgemeinen Einleitung in die vergleichende Anatomie, ausgehend von der Osteologie" ("First Outline of a General Introduction to Comparative Anatomy, Based on Osteology"). In his initial work on the intermaxillary bone, Goethe searched for a general design by observing the structural features common to all vertebrates. Through his study of morphological change, however, Goethe recognised that this approach fails to account for the various forms that vertebrates can take. The task is not simply to acknowledge *that* organic structure is fitted to a particular environment, but to grasp *how* organic structure exists in and by means of its environment. For example, instead of claiming "that a bull has been given horns so that he can butt", we should instead "try to discover how he might have developed the horns he uses for butting" (LA I 9:125). Thus Goethe sketched a new method for comparison that considers anatomical form as the result of a process by which the structure of living beings is "shaped from without as well as from within" (LA I 10:121). The goal is to extend analysis from the comparison of varieties against a standard, such as the human being, to the entire animal kingdom to derive a common type. The generality of the type insures that "no single animal can be established as a canon of comparison; no single one can be the pattern for the whole" (LA I 9:121). This does not mean that the type is a transcendent idea; rather, it is something "that must be established physiologically" (LA I 9:121). The type is the means to investigate the structures of all animals without having to refer back to the structure of the animals previously studied.

Goethe's language in "First Outline" is more epistemically restrained than his earlier texts. He states that his procedure for comparative anatomy begins with "experience [*Erfahrung*]", which "must first teach us what parts are common to all animals and how these parts differ" (LA I 9:121). From this general comparison, we can then search for an idea that governs the whole, one that "abstracts the general picture in a genetic way" (LA I 9:121). In this empirical procedure, the type serves as a hypothesis that "we may test quite adequately by applying the customary methods of comparison" (LA I 9:121). Once the hypothesis has been established, comparative anatomy can proceed via two levels of analysis. On the first level, the type enables one to consider the process of formation as limited by a particular structural plan. On the second level, it enables one to examine the enormous variety of animals as a graded series of organised forms. Holding these two levels together is the key to comparative anatomy, for their combination reveals that while "the structural range of nature is limited, because of the number of parts and their frequent modifications, the changes of shape become unlimited" (LA I 9:124).

The coupling of type and structural change marks an underappreciated shift in Goethe's morphological thinking.[5] Together, the two elements set

[5] An exception here is Bersier (2005), who attributes Goethe's shift to the influence of Kielmeyer's (1793) famous lecture at the Hohen Karlsschule. While I agree that the "First Outline" is deeply

comparative anatomy with the task of searching for the laws that govern the relation (*Verhältnis*) between type and environment. The type, Goethe states, must be "considered from the standpoint of how various natural forces affect it, and how it is likewise subject to the general laws of the external world" (LA I 9:126). For example, one can note that the neck and extremities are favoured in the giraffe, which has emphasised the reach of exterior limbs. The mole exhibits the reverse relation; the extremities are downplayed in order to prioritise its intensive relation to the soil. Such observations, Goethe states, give rise to a single law: "nothing can be added to one part without subtracting from another, and vice versa" (LA I 9:124).

Goethe's examination of the relation between the parts of an organic system draws from the programmes of comparative anatomy outlined by Johann Friedrich Blumenbach and Carl Friedrich Kielmeyer. In *Über den Bildungstrieb* (*On the Formative Drive*, 1781/9), Blumenbach (1789, 25) proposed a hierarchy of five life forces that account for the various organic functions, beginning with Haller's irritability and sensibility and extending to conception, nourishment and reproduction. Yet Blumenbach recognised that, on their own, the five life forces cannot account for the shape of organised beings. To include shape in the study of organic form, the life forces must be organised under a higher force that manifests quasi-agential characteristics, which he described as the formative drive (*Bildungstrieb*). In contrast to the five life forces, which concern predefined capacities, the formative drive plays a coordinating role. It is not a vital power, but operates as the placeholder for an unknown cause. Building on Haller's analogical extension of Newtonian gravity, Blumenbach (1789, 25–26) states that the formative drive "should serve no more and no less than to signify a force [*Kraft*] whose constant effect is recognised from experience, and whose *cause*, like the causes of the aforementioned widely recognised natural powers, is for us a *qualitas occulta*".

In a lecture given at the Hohen Karlsschule in 1793, *Über die Verhältniße der organischen Kräfte* (*On the Relations of the Organic Forces*),

indebted to Kielmeyer, in what follows I suggest that Goethe sees beyond Kielmeyer's prolegomena to a future comparative anatomy that examines the relation between the organic forces as the distribution of the *Bildungstrieb*.

Kielmeyer proposed a new method for physiology that examines the distribution of form across the animal kingdom in terms of the relation between the five life forces. Kielmeyer was aware of Kant's interpretation of Blumenbach in the third *Critique*, which denied that Newtonian gravity could serve as an analogue for the formative drive. In Kant's view, our own purposiveness as agents provides the analogue, thus removing the formative drive from the causality of nature. Yet in contrast to Kant, Kielmeyer rejects the formative drive altogether. Instead, he removes Blumenbach's account of the five life forces from a Newtonian frame and proposes a revised programme in which the living world is examined as a biodynamic system of interactions. The organisation of the life forces is governed by the "law of compensation [*Kompensationsgesetz*]", by which the life forces are proportioned in relation to an organism's need under particular environmental conditions (Kielmeyer 1793, 35–36).[6] The law of compensation is not grounded in a force, for nothing is directing the interplay of the forces from without. In fact, Kielmeyer does not tell us in what sense it is a law at all.[7]

Goethe's system of dynamic relations has clear resonances with Kielmeyer's study of the relation between the life forces. His law of exchange is a direct paraphrase of Kielmeyer's law of compensation (see Bersier 2005, 25–26). Yet in contrast to Kielmeyer, Goethe retained Blumenbach's formative drive as the force that coordinates the other functions. Here the singularity of Goethe's programme is apparent. In place of Blumenbach's Newtonian analogy, and in contrast to Kielmeyer's rejection of any coordinating force, Goethe redeploys the economic analogy originally set out by Linnaeus. When placed within a self-regulating system of exchange, the formative drive becomes the distributive principle that, in response to the external forces of an environment, emphasises certain structural features at the expense of others:

> Here are the boundaries of animal nature, in which the formative force [*die bildende Kraft*] appears to move in the most wonderful, almost capricious way, without being in the least capable of breaking the circle or leaping out of it. The formative drive [*Bildungstrieb*] is here set as the ruler of a bounded

[6] Kielmeyer presents the law as follows: "the more one of these forces on one side is cultivated, the more they are neglected on the other".

[7] For an analysis of Kielmeyer's relation to Newtonian physiology, see Cooper (2020).

> and yet well-supplied kingdom. The rubric of its budget, in which its expenses are to be distributed, are prescribed to it, and yet what it wants to give to each is, to a certain extent, free. If it wants to let one have more, it may do so, but not without taking it from another. Thus nature can never fall into debt, even less go bankrupt. (LA I 9:125)

By redeploying Linnaeus' economic analogy, Goethe is able to maintain both the freedom of and the constraint on the formative drive. On the one hand, the formative drive is entirely free to allocate resources as it pleases, giving rise to new shapes and distributions without a predetermined goal. On the other hand, it is constrained to the structural plan, meaning that it distributes the available resources according to the "idea of a household-like give and take", increasing some parts of the organism in complexity while decreasing others, rendering them more efficient for the propagation of the organism (LA I 9:126) (see Lenoir 1987, 24). The law of compensation ensures that the development of any one part is accompanied by a depreciation of others, meaning that some organisms become less complex. The result is a dynamic relation between inner and outer:

> If one inquires into the causes that bring such a manifold of determinations to light, then we answer above all: the animal is formed by external conditions for external conditions; thus its inner perfection and its external purposiveness. (LA I 9:126)

Here Goethe works closely with Kant's idea that the organism can be considered according to both its inner perfection and external purposiveness. Yet rejecting the limitations that Kant placed on teleological judgement, Goethe contends that both inner and outer purposiveness can be understood as two modes of reflecting on a single cause. When an organism is examined according to its functional parts, their arrangement is sealed by and for external conditions. When the organism is examined according to its inner perfection, the arrangement of the parts can be explained according to the dynamic relation between type and environment. On land, for instance, the exterior organs used for motion are arranged in proportion to the body. Take the snake: the extreme length of its body explains why it has no additional organs. As soon as we find these additional organs in another formation (*Bildung*), say in the lizard, the body is much shorter. In the frog, the long legs mean that the form is contracted even further, and the unformed toad is given shape according to this law of compensation. In the conditions of the ocean, the body of the fish takes up the maximal proportion of its shape "due to the laws of the elements" (LA I 9:127). According to "the laws of the organic type", however, the swelling of the body is compensated by the contraction of the outer organs (LA I 9:127). The type which develops in the air has a drier interior, meaning that a lean form emerges in the bird so that its formative force (*bildende Kraft*) has sufficient resources to mould the skeletal structure and feathers which enable it to fly. "So the eagle is formed both through the air and by the air, through the high mountains and by the high mountains", meaning that the shape of the parts, and their relation to one another, "are determined by what the creature needs to remain alive" (LA I 9:127). The swan and the duck thus reveal their inclination to the water by their shape. When viewed within nature understood as a biodynamic system of exchange, "the effect of the climate, the height of the mountain, the heat and the cold, together with the effects of the water and the common air, becomes extremely powerful for the formation [*Bildung*] of animals" (LA I 9:127).

Goethe's "First Outline" proposes a radical shift in comparative anatomy. Instead of asking *why* certain features are present in a species, it directs the comparative anatomist to ask *how* the formative force finds room for those features in its budget. To find out how a particular feature came to be there, the comparative anatomist searches for the adaptation in other features that may have prompted the formative force to make certain compensations. The productive tension between type and metamorphosis ensures that while a single organism is constrained by its type, its household economy is not closed. The parts are proportioned in response to external conditions in order to balance the requirements of the structural plan.

Goethe revised "First Outline" the following year in a text entitled "Vorträge über die drei ersten Kapitel des Entwurfs einer allgemeinen Einleitung in die vergleichende Anatomie, ausgehend von der Ostelogie" ("Lectures on the First Three Chapters on the Outline of a General Introduction to Comparative Anatomy, Based on Osteology", 1796). He expands on the theoretical section of "First Outline", emphasising the higher investigative stance made available by the economic analogy:

> Thus we would seem to be justified in claiming without question, that perfect organic natures, including fish, amphibians, birds, mammals, and, at the pinnacle of the last, the human being, are all formed according to an archetype [*Urbilde*] that deviates more or less only in its very stable parts, and still develops and transforms itself day by day through propagation. (LA I 9:198)

The view afforded from this higher stance bears striking similarities with the system of relations sketched in Kielmeyer's lecture. While there is an order to nature's incessant change in the form of a set of constraints that unite the more differentiated animals, the actual change cannot be understood according to a preestablished end. Exactly where the formative drive allocates the finite resources of its budget is underdetermined by external forces. The goal of comparative anatomy is thus akin to art history: to capture the "certain and unequivocal genius of productive nature" (LA I 9:200). The human being is the pinnacle of the animal kingdom only to the extent that it is the most differentiated and complex organisation. Yet it is neither the goal of the type's morphology nor even a necessary step along nature's ascent. Transformation is not a linear process, and perfection exists in relation to the archetype. Since it is the archetype that defines the possible variations, no individual case, such as a particular species or even an actual ancestor, can provide a model:

> If one has but grasped the idea of this type, one will be fully aware of how impossible it is to declare a single genus [*Gattung*] to be the canon. The singular can never be the model for the whole, and so we must not search for the pattern for all in the singular. The classes [*Klassen*], genera [*Gattungen*], species [*Arten*] and individuals [*Individuen*] stand as cases to

the law; they are contained within it, but they do not contain it and cannot give it. (LA I 9:199)

The economic distribution of resources transposes an externally ordered hierarchy into a system of biodynamic exchange, wherein the unity of the system is a natural achievement. In such a system, a genus is not a logical category but rather a principle that lies behind every instance, one that can only be distilled by observing the process of formation. Any particular state is derivative from the continual process of formation, such that each isolated manifestation gives expression to a developmental path. The distribution of the available resources determines the degree of sophistication of living organisms, meaning that the human being, despite its "high degree of organic perfection, [...] stands least of all as the measure for all the other less perfect animals" (LA I 9:199). Here Goethe's thought is continuous with his essay on Kant: "the phenomena must form a series, or rather, overlap; thus they give the scientist a picture of some organization by which the inner life of the phenomena become manifest as a whole" (LA I 9:90).

5 The Formative Drive

In the years following "First Outline", Goethe continued to work on the methodology for a new system of comparative anatomy based on morphology. Yet no substantive work was forthcoming. At several points during the following decade he expressed his intent to collect his writings into a single volume, yet this did not eventuate until *Zur Morphologie* (*On Morphology*) in 1817. In the prefatory essay, Goethe presents morphology as an experimental science that begins with observation and then searches beyond every observation for the principle behind the development in actual living beings (ZM I xx–xxxii). The key to comparative anatomy is to balance the need for a structural plan with plasticity, such that the plan itself can alter in the course of the sequence. Here it is clear that any ontological kind of essentialism has vanished from Goethe's thought (see Richards 2017, 232–233). The plasticity of the structural plan is what separates organic from inorganic nature. While physical matter reiterates its expressions unchangingly, the shifting relation of the parts brought about by the formative drive means that organic life incessantly alters its shape and thus also its needs. In every change something comes to be that did not previously exist, yet morphology is nevertheless governed by an unchanging law.

In "Bildungstrieb" ("Formative Drive", 1810), a condensed essay published in *Zur Morphologie*, Goethe presents his morphology as the completion of a progression that begins with Haller, continues through Wolff and culminates in Blumenbach. At first glance, this brief history of physiology seems to collapse Haller's evolutionary conception of development into Blumenbach's epigenetic account of generation. Yet Goethe's point is that as far as comparative anatomy is concerned, both follow a Newtonian model in which forces are considered to be external to the bodies they move. When conceived of through the Linnaean analogy, the formative drive decouples the study of generation from the limitations of Newtonian force, enabling the natural historian to view the natural system as a self-regulating economy.

Goethe begins the essay by identifying Haller as the beginning of a progressive movement away from the preoccupation with external purposes to account for organic form. By extending the Newtonian analogy to physiological movement, Haller identified two life forces, sensibility and irritability, which stood as placeholders for the properties of muscle fibres; irritability stands for the property of insensitive muscle fibres, which respond immediately to stimulation, and sensibility for the property of sensitive fibres, which alerts the organism to a change in external conditions (Haller 1936, 658). While the life forces made it possible to refer to the inner properties of organic fibres, Haller denied that they can account for morphological change. Rather, the stimulation of muscle fibres causes a preestablished germ to unfold and to grow (Haller 1758, II 172). In contrast, Wolff advocated an epigenetic view in which an organic kind of matter, endowed with an essential force (*vis essentialis*), sustains living creatures. In Goethe's terms, Wolff's *vis essentialis* "adapts itself to all that is produced, and is thus elevated in its own right to the level of a producer" (LA I 9:99).

Despite the differences between Haller's conception of evolution and Wolff's epigenesis, Goethe claims that both remained tethered to Newton's material conception of force. Thus framed, "the word 'force' means something purely physical, even mechanical; the question of which organism is to arise out of that substance remains obscure and insoluble" (LA I 9:99). Blumenbach achieved the "ultimate refinement" of this idea, for he "anthropomorphized the phrasing of the riddle and called the object of discussion a *nisus formitavis*, a drive [*Trieb*], a surge of action [*eine heftige Tätigkeit*] which was supposed to cause the formation [*Bildung*]" (LA I 9:99). Blumenbach thereby removed the need to assume a preestablished form, and thus an external purpose, for his concept of the drive is not a predetermined capacity but rather a principle of formation. Yet Blumenbach too remained tethered to Haller's Newtonian analogy. The formative drive ultimately serves as a placeholder for the unknown cause of the formative process, which is somehow material *and* purposive.

Goethe's claim is that by examining the physiological faculties through an analogy with Newtonian force, and by subsuming the faculties under the formative drive, Haller, Wolff and Blumenbach began to free a science of morphological change from the constraints of physicotheology. Haller recognised that to develop a science of physiological movement, one must presuppose an action that is prior to it. Wolff demonstrated that to form a concept of action, we require a suitable material on which to act. If this action is personified, then we are led to the concept of "a god, as a creator and sustainer, whom we are constrained to worship, honour and praise" (LA I 9:100). If the action is anthropomorphized, however, and seen as "coexisting with the underlying material", a new framework for examining organic form arises (LA I 9:100). The development from Haller to Blumenbach thus shows that "any theory of accommodation and adaptation will have to presuppose something which adapts and something to which it adapts" (LA I 9:100). Blumenbach's anthropomorphism, which recasts the life force as a drive, sets anatomy on a new course, for it transforms the preestablished form of evolution into an activity that contains the spontaneity and adaptation of epigenesis. Goethe contends that this development is completed in the concept of metamorphosis, which enables one to maintain "the unity and freedom of its formative drive [*Bildungstrieb*]" (LA I 9:100). Metamorphosis does not deal with fixed shapes (*Gestalten*) but with progressive formation (*Bildung*), wherein developmental regularities in each individual organism express its type, revealing the regularities across multiple types. The comparative anatomist can thus work between various levels of analysis, first noting the common principle of a species, then the common principle of a genus, and then a class, all the way up to the animal kingdom, and even to nature as a whole. From each level she can observe the distributive activity of the formative drive, which, bound by the law of compensation, is free to distribute the finite quantity of resources wherever it will.

6 Conclusion

In this chapter I have argued that Goethe's economic analogy opens a new framework for examining nature as a large-scale system of exchange. In "First Outline", Goethe weaves the idiosyncratic typology of his Italian journey together with the major developments in comparative anatomy of his time, demonstrating how a structural plan is required in a system of endless mutability. He recognised that Haller's extension of Newton's forces to physiological properties removed the study of organic form the domain of physicotheology, and he saw that Blumenbach's formative drive offered a new vantage on the generative process in which the structural plan and organic matter were considered together. Yet in Goethe's view, the Newtonian analogy restricted anatomy to the comparison of organic function, thereby removing the perceptible and incessant flux of organic form from scientific examination. He redeployed the Linnaean idea of a natural economy as an alternative analogy by which to search for a principle that governs the distribution of resources responsible for the creative change of developmental pathways: the formative drive.

Goethe's proposal is, of course, provisional. It does not explain how the agential properties of the formative drive differ from our own agency as creators, and it remains ambiguous, even in his later work on morphology, whether the underlying pattern is imposed by our minds or is the template followed by an independently creative nature. Nevertheless, his programme of comparative anatomy is not incompatible with evolutionary theory. While he remained ignorant of the mechanics of generation, Goethe proposed an observational method that does not require exhaustive knowledge of mechanical connections, but opens a higher view of nature as a system of dynamic exchange organised according to a law that cannot be reduced to mechanics. This is precisely what captured Darwin's attention as he prepared *The Origin of the Species*. Darwin (1959, 295) explains that he was inspired by Goethe's "law of compensation", which states that "in order to spend on one side, nature is forced to economise on the other side". This reference to Goethe's law occurs in his examination of the laws of variation, where Darwin suggests that the external pressures placed on a plant select certain characteristics at the expense of others. Clearly Darwin found in Goethe's economic analogy a way to examine the phenomenology of change in contexts where external pressures determine the expression of internal form. His uptake of the self-regulating conception of nature's economy gives us reason to conclude that Goethe's role in the formation of evolutionary science was more extensive than modern historians of biology have been ready to accept.

References

Benn, Gottfried. 1949. *Goethe und die Naturwissenschaften*. Zürich: Arche.

Bersier, Gabrielle. 2005. Visualizing Carl Friedrich Kielmeyer's Organic Forces: Goethe's Morphology on the Threshold of Evolution. *Monatshefte* 97 (1): 18–32.

Blumenbach, Johann Friedrich. 1789. *Über den Bildungstrieb*. 2. Aufl. Göttingen: Johann Dieterich.

Boyle, Nicholas. 2000. *Goethe: The Poet and the Age*. 2 Vols. Oxford: Oxford University Press.

Brady, Ronald. 1987. Form and Cause in Goethe's Morphology. In *Goethe and the Sciences: A Reappraisal*. Edited by Frederick Amrine, Francis J. Zucker, and Harvey Wheeler, 257–300. Dordrecht: D. Reidel.

Cooper, Andrew. 2020. Force and Law in Kielmeyer's 1793 speech. In *Kielmeyer and the Organic World*. Edited by Daniel Whistler and Lydia Azadpour, 81–98. London: Bloomsbury.

Darwin, Charles. 1959. *The Origin of Species by Charles Darwin, a Variorum Text*. Edited by Morse Peckham. Philadelphia: University of Pennsylvania Press.

Goethe, Johann Wolfgang. 1817. *Zur Morphologie*. 2 Vols. Stuttgart and Tübingen: J. G. Cotta.

———. 1888. *Goethes Werke* (Weimar Ausgabe). 133 Vols. Weimar: Hermann Böhlau.

———. 1947. *Goethe: Die Schriften zur Naturwissenschaft* (Leopoldina Ausgabe). 21 Vols. Edited by Dorothea Kuhn. Weimar: Hermann Böhlau Nachfolger.

von Haller, Albrecht. 1758. *Sur la formation du coeur dans le poulet*. Lausanne: Bousquet.

———. 1936. A Dissertation on the Sensible and Irritable Parts of Animals. *Bulletin of the History of Medicine* 4 (8): 651–691.

von Helmholtz, Hermann. 1865. Über Goethes naturwissenschaftliche Arbeiten. Vortrag gehalten im Frühling 1853 in der deutschen Gesellschaft zu Königsberg. In *Populäre wissenschaftliche Vorträge*, vol. 1, 31–55. Braunschweig: Friedrich Vieweg und Sohn.

Holland, Jocelyn. 2009. *German Romanticism and Science: The Procreative Poetics of Goethe, Novalis, and Ritter*. London: Routledge.

Kielmeyer, Carl Friedrich. 1793/1993. *Über die Verhältniße der organischen Kräfte unter einander in der Reihe der verschiedenen Organisationen, die Gesetze und Folgen dieser Verhältniße*. Facsimile reproduction edited by Kai Torsten Kanz. Marburg: Basilisken.

Kuhn, Dorothea. 1987. Goethe and Theories of Development. In *Goethe and the Sciences: A Reappraisal*. Edited by Frederick Amrine, Francis J. Zucker, and Harvey Wheeler, 3–16. Dordrecht: D. Reidel.

Larson, James. 1967. Goethe and Linnaeus. *Journal of the History of Ideas* 28 (4): 590–596.

Lenoir, Timothy. 1987. The Eternal Laws of Form: Morphotypes and the Conditions of Existence in Goethe's Biological Thought. In *Goethe and the Sciences: A Reappraisal*. Edited by Frederick Amrine, Francis J. Zucker, and Harvey Wheeler, 17–28. Dordrecht: D. Reidel.

Linnaeus, Carl, and Isaac J. Biberg. 1749. *Oeconomia Naturae*. Uppsala: Medelpadus.

Mayr, Ernst. 2000. *The Growth of Biological Thought: Diversity, Evolution, and Inheritance*. Cambridge, MA: Harvard University Press.

Richards, Robert. 2002. *The Romantic Conception of Life: Science and Philosophy in the Age of Goethe*. Chicago: University of Chicago Press.

———. 2017. Did Goethe and Schelling Endorse Species Evolution? In *Marking Time: Romanticism and Evolution*. Edited by Joel Faflak, 219–238. Berlin: De Gruyter.

Wells, George. 1978. *Goethe and the Development of Science, 1750–1900*. Alphen aan den Rijn: Sijthoff & Noordhoff.

Whewell, William. 1847. *History of the Inductive Sciences*. 3 Vols. London: Parker and Son.

6

"Wie die Triebe, so der Sinn; und wie der Sinn, so die Triebe": Jacobi on Reason as a Form of Life

George di Giovanni

1 Jacobi and the Enlightenment

Jacobi had no doubt about the one factor that had motivated him unchanged throughout his long philosophical career, despite the ever shifting historical circumstance in which that career had played itself out.[1] It was the belief that the world is the creation of a transcendent

[1] I have given an account of this career, placing it in historical context, in di Giovanni (1994 & 2009), an introductory monograph included in Jacobi (1994 & 2009). This last is hitherto cited as MPW, followed by a page number. The present chapter is the twin of another ("Jacobi and the Poets") that shows how, according to Jacobi, speculative nihilism has an existential counterpart in a type of personality which is just as destructive at the level of lived human relations as metaphysical monism is at the conceptual. The two papers differ in intent. The present is concerned with the early Jacobi's speculative theory of experience; the other, with Jacobi's two novels, *Allwill* and *Woldlemar*. Nonetheless, both presuppose Jacobi's understanding of Spinoza, of which I give an account in a first stage of both papers. In this, the two papers reflect each other, and in some places the present follows the earlier *verbatim*. It would have been disingenuous on my part to cover up an identical reading of Jacobi's text, equally relevant to both papers, with superficial differences in wording.

G. di Giovanni (✉)
McGill University, Montreal, QC, Canada
e-mail: george.di_giovanni@mcgill.ca

M. Kisner, J. Noller (eds.), *The Concept of Drive in Classical German Philosophy*,
https://doi.org/10.1007/978-3-030-84160-7_6

God, free and intelligent, such as human individuals are able to relate to as person to person. He had held this belief in opposition to any view of reality that would rather make of freedom and intelligence, whether human or divine, only surface phenomena of an otherwise originally amorphous yet self-evolving nature. In 1815, in the Introduction to his planned Collected Works, he reiterated this personalist manifesto, concluding with the words: "So I spoke at the beginning. I end now as I began" (MPW 590). So he did. Nonetheless, while Jacobi had clearly and repeatedly asserted his belief throughout his life, unchanged in a variety of different contexts, he had not uniformly defended it. His grounds of defence shifted. The reason was a tension in him that pulled him in different directions. It was a tension characteristic of the late Enlightenment that also affected him personally and gave his motivating belief its special quality.

Jacobi himself was aware of it. On the one hand, there was the culture of the *Herzensmensch*, of the pietist who put his trust in the sentiments of the heart, and in religious beliefs derived from tradition. Jacobi squarely belonged to it. The young Goethe had occasion to poke fun at the sentimentalism of Fritz and Georg, the two Jacobi brothers (Goethe 1774, 1777). On the other hand, there also was the culture of reason that instead put its trust in the power of the concept. Jacobi was strongly drawn in this direction as well. In his own days, he was widely considered a fideist, and in the literature, past and current, he has generally been portrayed as a counter-Enlightenment figure.[2] But in fact, Jacobi never opposed reason as such. What he wanted was a rationality that would rather respond to the heart-felt needs of the pietist. In dispute with Mendelssohn in 1775, with his penchant for the *bon mot*, he came up with a formula that was intended to define his position, but in fact rather gave voice to the tension that underlay it. As he said to Mendelssohn "Wie die Triebe, so der Sinn; und wie der Sinn, so die Triebe" or, loosely yet in context accurately translated, "As the heart, so the mind; as the mind, so the heart" (MPW 237). Balancing the two sides of the

[2] For a now classic portrayal of Jacobi in this guise, cf. Beiser (1987). For a view that makes Jacobi very much part of the Enlightenment, cf. di Giovanni (2005), especially 77–91.

Enlightenment, and of his personality no less, was Jacobi's driving force. It was his problem as well.

Whether the balancing was possible on Jacobi's terms is the issue. But it is worth noting at the outset that as close a friend of Jacobi as Hamann faulted him for sinning on the side of reason. At the time of the Mendelssohn dispute, Hamann was sympathetic to the cause of Jacobi, whom he was fond of calling "my Jonathan." Like Jacobi, Hamann also resented the fuss the Berliners were making around this Jew Mendelssohn, commemorating him upon his sudden death in the course of the dispute. In this, Hamann, like Jacobi, was showing the dark side of his personality. But, contrarian by nature as he was, Hamann never hesitated to needle Jacobi for what he thought was his propensity to make undue excursions into the domain of his adversaries. Engaging philosophically with Mendelssohn was a case in point.[3] Such excursions only succeeded in empowering reason. "Your theory [of truth]," he complained to Jacobi, "is truly a piecemeal assortment of philosophical and human authorities." And he continued, in language nowadays terribly incorrect politically, "Fail to feel this, dear Jonathan, and you'll go the way of women who turn from street walkers into church pillars" (Hamann 1955–1979, Vol. VII, Letter #1060, 175).

We might not want to put much weight on Hamann's witness. His sacramental sense of reality made him temperamentally incapable of appreciating anything speculative. But there is also Schelling's witness. In his Munich Lectures of 1833–1834, Schelling gave an even-handed review of Jacobi's work that still stands up to critical judgement (Schelling 1994, 164ff.).[4] In his estimate, no philosopher had ever conceded as much as Jacobi had to the rationalism of the Enlightenment. He had disarmed himself before it by granting that its knowledge, based on the abstractions of reason, was the only one scientifically possible. But this knowledge pre-empted the possibility of the personalism that was close to

[3] There was a further reason for why, according to Hamann, Jacobi should not have engaged in discussion with Mendelssohn. As he wrote to him: "Certainly you [Jacobi] are to be blamed, and this is your unacknowledged guilt, that you have sought and presupposed *truth* in a Jew, a natural enemy of it" (Hamann 1955–1979, Vol. VI, Letter #939, p. 299). Translations of Hamann's texts are mine.

[4] Cf. also, di Giovanni (2021), Chapter 4.1.

his heart. Jacobi, accordingly, counterposed to it another kind of knowledge, one that could not be demonstrated argumentatively but was nonetheless self-validating, for it responded to a knower's deeply rooted subjective needs. As contrasted with scientific knowledge, this was a non-knowledge that nonetheless carried conviction. Jacobi called it "faith," a term for which Schelling had no objection in principle, since he agreed with Jacobi that all conviction is at origin motivated by naturally and historically determined dispositions. But since one is simply born into these, their witness to the truth, according to Schelling, is at first accepted spontaneously; consequently also uncritically—in other words, in faith. In order for science proper to come to the scene, there is the further need of the work of the understanding which, by virtue of conceptual reflection, transforms into rationally established positions otherwise immediately held beliefs. Schelling understood Jacobi as initially holding precisely this position; to that extent, also as presaging him. In the review, he pointed out, quite rightly, that Jacobi had explicitly said that knowledge is historical in nature, exactly his view.[5]

But, Schelling went on to say, in the Enlightenment climate of the day, Jacobi had been attacked for the use of the term. He had been accused of fideism and therefore of irrationalism. And, bowing to this pressure, Jacobi had hit upon the stratagem, which Schelling found disingenuous, of replacing "belief" with "reason," where by "reason"; he, however, meant immediate "feeling." The move was disastrous. For by feeling, now baptized as reason, Jacobi meant an immediate and therefore vague apprehension of reality that was just as empty of content as were the abstractions of the philosophers, the same abstractions because of which he had earlier accused Spinoza, and the rationalism that he represented, of nihilism. Jacobi even went so far, Schelling complained, as to ask his readers, when preparing his collected works late in life, to read "understanding" where earlier he had "reason." He (Jacobi) now realized that the understanding was responsible for the abstractions for which he had earlier blamed reason. To the negativity of the latter, the readers had rather

[5] "They had no philosophy, or rather, their philosophy was history" (MPW 238). Jacobi is referring to two Spartans who, as the story is recounted by Herodotus, refused to bow before Xerxes, the King of Persia, thus exposing themselves to death.

to oppose the positivity of a science based on the immediate intuition of reality, such as reason alone, now however understood as "feeling," could yield. The net result was that the positive science Jacobi now proposed was a distorted counterpart of the rationalistic metaphysics he had originally opposed, and still opposed. Thus, where Jacobi had disarmed himself before it at a first stage, in a second, just like it, he relied on the vagueness of intuition to convey the impression that one reached beyond the senses, while in fact remaining bound to them.

Schelling was referring to the already-mentioned Jacobi's Introduction, and it is true that after 1800 Jacobi was following Jakob Fries, assigning the higher truths of Spirit to the direct intuition of reason (allegedly Platonic in nature) while reserving for the conceptual work of the understanding the task of organizing the otherwise scattered content of sense-experience. In this, however, he deprived the sciences of nature of all speculative value, not a move that Schelling would have appreciated. We shall return to this late Jacobi. Of interest at the moment, however, is the early Jacobi, the one who Schelling rightly believed had anticipated him (and Hegel as well, one should add). The two relevant texts are the *Spinoza-Letters* and the Dialogue *David Hume*—this last in its first edition of 1787, before Jacobi, in the second edition of 1815, obfuscated it with tendentious notes, as Schelling complained.[6]

2 Jacobi and Spinoza

Jacobi knew his Spinoza. He had first-hand acquaintance with his texts. He understood that on Spinoza's definition of substance there is no room for a creation *ex nihilo*; indeed, even for the coming-to-be of anything new, however limited in scope this "new" would be. Creation, and becoming in general, require prior nothingness, and this is a circumstance that *being* excludes. Whether one calls it God or, as Spinoza did, Substance, being is per se already *all there*. One can still distinguish it from the finite things we experience in space/time. But, that such things stand on their

[6] The English translations of both texts, from original editions, and, in the case of the *David Hume*, with notes indicating the changes introduced in 1815, are in MPW.

own, that is, that they *exist*, requires that they *not be* something else (*omni determinatio negatio*). It follows that what they presumably are in themselves is definable only with reference to precisely this "something else," that is, on the basis of external factors: in effect, therefore, as *not* anything in themselves. This is a contradiction. When it comes to defining them *in themselves*, one must therefore fall back upon being *per* se. But the latter, just as Kant says of space with respect to spatial things,[7] is already wholly present in each of them. Any presumed difference between them, or between them and being *per se*, is thus rendered moot. In essence, things are only the illusionary modes of one substance existing per se and *a se*. They are semblances of being rather than being.

This, in brief, is how Jacobi expounded Spinoza's position in conversation with Lessing in 1780, and eventually in correspondence with Mendelssohn (MPW 187–188; 217–220). Hypothetically, Jacobi accepted it, because, on the ideal of explanation that motivated the metaphysics of the day, it was in his estimate the only rigorously consequential. Not explanation, however, but its consequences at the personal level of existence, or what it would be like for one to live in a universe if it truly were as Spinoza described it, or one effectively believed it to be such, was the concern that motivated Jacobi. The consequences were clear. It meant that despite one's undeniable feeling of being an agent among other agents relating to them precisely as agents, that is, despite what one's heart undoubtedly says, one is in truth only an observer: of the other presumed agents and of oneself no less. The keyword here is "observer." It meant standing at a distance from all things, including oneself, without a definite *point d'appui* from which to observe, in truth only registering lines of events that anonymously move forwards, driven from behind. In Spinoza's universe, if one were to attach a determinate name to any of these events, the connection would be only external, with no internal justification on the part of the event itself (MPW 189ff.; 194–195).[8]

But that was not for Jacobi, above all because Spinoza's universe foreclosed the possibility of the person-to-person relation with God that his

[7] The comparison between Spinoza and Kant is Jacobi's. Cf. MPW 218, Jacobi's long footnote 30.
[8] In the *David Hume*, referring to the *Spinoza-Letters*, Jacobi says that "the faculty of thought is always a spectator, and cannot ever be a source of external action" (MPW 291).

pietistic leanings demanded. He knew that there was no point arguing against the philosophers. On their ideal of explanation, Spinozism was conceptually unimpeachable. But Jacobi was also convinced that the ideal was in fact parasitic on a deeper source of truth which is innate in us. Hence, although there was no point arguing against the philosophers, one could nonetheless exhort them to alter their attitude regarding truth. Rebuffed by Lessing, Jacobi simply invited him to place himself on that "elastic spot" (presumably his inner self, the heart) whence to perform the jump (a somersault, in effect) that would right his position and have him walk on his feet where before, like all philosophers, he had walked on his head (MPW 189; 195). Significant is that, in performing the jump and landing on his feet, Lessing would in fact be walking alongside the *Herzensmensch.*

Lessing could not be blamed if he wondered at one point whether he was in the presence of one of the many so called *Schwärmer*, the religious enthusiasts of the day. Jacobi was indeed drawing from the other side of his personality, the one that resonated with the culture of the *Herzensmensch.* Later, when in dispute with Mendelssohn, ever the philosopher *malgré soi* as Hamann complained, he summed up his position with the already-mentioned formula (MPW 237; 230ff.). As Jacobi explained, we would not seek the truth unless we were already motivated by it; unless it were already with us, in other words, but unconsciously at origin, organically given at the primary level of bodily instinct, as *Trieb.*[9] The *salto mortale* Jacobi had recommended to Lessing was no leap into the unknown. On the contrary, it was a matter of actively retrieving a possession already secured by the heart, consciously making it one's own, thereby giving rise to *Sinn.* It followed, as Jacobi spelled out for Mendelssohn, that philosophy can only be descriptive. It was the felt story of how one comes to the convictions that shape one's identity. It has to be historical, in other words (MPW 239). It also has to have social relevance, because—as Jacobi said in the *Spinoza-Letters* and was never tired to repeat—*there is no I without a Thou* (MPW 231). Or again, as Jacobi also said, a living philosophy can only grow out of the life of a

[9] The "bodily" character of *Trieb* comes through clearly only in the *David Hume*, as we shall see further.

people. It's not philosophy that determines the institutions of a people but the people's lifestyle that rather determines its philosophy (MPW 240ff.).

This was the side of Jacobi for which, as Schelling later acknowledged, he had indeed presaged him. But, in retrospect, Schelling was crediting him with too much. He attributed to him the negative work of the understanding, which is precisely what Jacobi's formula would have required in order for the heart/mind reconciliation that it proclaimed to be anything serious. The two sides needed clear distinction. As the formula stood, however, it begged the question of how the attained truth can at once be *felt* as *Trieb*, yet at the same time sufficiently transcend the latter so as to be itself, as *truth*, more than just a mere product of nature or an accidental vicissitude of history. It begged the question of how *Sinn*, as it emerges, qualifies *Trieb de novo*. In his exchange with Mendelssohn, Jacobi himself unwittingly gave witness to the flaw. He called *Glaube* the historical, quasi-instinctual disposition to accept a truth before the philosophers give voice to it reflectively (MPW 230–231). As we said, Schelling had no objection to this usage. In German *Glaube* carries both the not necessarily religious meaning of "belief" and the definitely religious one of "faith." He also called *Offenbarung* the truth's capacity to command assent by its very presence (MPW 231). To this also Schelling would not have objected. Truth—any truth—is inherently self-revelatory: it is *Offenbarung* or revelation, an unveiling, ἀλήθεια. But in context Jacobi was using the terms with unmistakably religious connotations. The concluding part of the *Spinoza-Letters* reads like a prolix exercise in pietist effusion in the spirit of Lavater. In saying to Mendelssohn, "Dear Mendelssohn, we are all born in the faith, and we must remain in the faith, just as we are all born in society, and must remain in society" (MPW 230), Jacobi was not just making a possibly legitimate conceptual point; he was intimating that Mendelssohn could not understand him because he was a Jew. He had missed the historical moment of Christ's revelation.

The flaw in Jacobi's formula, which was also the flaw of the *Herzensmensch*'s culture and of Jacobi's persona no less, was that it had no self-limiting principle, no effective internally negative principle. It lacked the labour of the understanding. In a letter to Kant, Mendelssohn said that Jacobi's Spinoza book was like a monster with Goethe as head,

Spinoza for body, and Lavater for feet (Kant 1999, 230; October 16, 1785). Kant, incidentally, reported the image to Hamann in strict confidence, but Hamann could not resist passing it on to Jacobi attributing it to vague sources. Jacobi was not pleased. Yet the image was a fair illustration of the confusion of which his unresolved inner tension made him the prey.

3 Jacobi and Hume

It is unlikely that in 1780, at the time of Jacobi's visit to Wolfenbüttel where Lessing resided, Kant was in Jacobi's or Lessing's mind. Goethe and his Promethean view of humankind which Jacobi considered pantheist were the issue. Nor was Kant at the centre of Jacobi's subsequent dispute with Mendelssohn,[10] even though by that time all those whether directly or indirectly involved in it were definitely aware of his presence. We learn from Jacobi's autobiographical notes that he had always been sympathetic to Kant (MPW 278ff.). As of 1785 he could well have believed that in fundamental points his philosophical agenda was not unlike Kant's. Indeed, the similarities were even striking. Like the Kant of the *Critique of Reason*, for instance, Jacobi believed that philosophy should be done from the standpoint of one whose *point d'appui* lies in experience itself, with no privileged access to it from outside. It therefore followed, for Kant no less than for Jacobi, that subjectivity is an essential factor in the determination of the objective content of experience; that the knowledge of this content is therefore inevitably descriptive and historical—"phenomenological," in one word. Like Kant, moreover, Jacobi believed that existence cannot be argued for on the basis of reason's abstractions—in fact, it cannot be argued for at all. It was on this point especially that he had originally discovered his affinity with the still precritical Kant.[11]

---

[10] Kant is mentioned in the *Spinoza-Letters* in one footnote. MPW 219, footnote 30.

[11] See Jacobi's reaction, as described by him, to Kant's *Enquiries Concerning the Principles of Natural Theology and Morality* (1764) and *The Only Possible Basis for a Demonstration of the Existence of God* (1763). Kant's privileging of actuality over possibility is the point that especially attracted Jacobi (MPW 281–282; 284–285).

The difference, of course, is that Kant had the internal limiting principle, which Jacobi lacked, for distinguishing between the accidentally subjective and the subjective which is rather the medium where objectivity is established in the first instance. This was the a priori for which Jacobi bitterly criticized Kant in a Supplement to the first edition of the *David Hume* (MPW 330–338). He criticized him on other grounds as well that became canonical in the reception of Kant's *Critique*. All this is well known. Not often noticed, however, is the criticism he had already indirectly levelled against him in the body of the Dialogue itself. He had suggested what amounted to an alternative to Kant's a priori—one, however, for which there was no need to resort to conceptual abstractions. It did not run the risk, therefore, of lapsing into subjectivism and dogmatism as Kant, in Jacobi's opinion, had done.

To make his case, Jacobi visited in the Dialogue the most disparate of places: the aftermath of the dispute with Mendelssohn; Hume's idealism; reminiscences of Jacobi's early life; Jacobi's early reactions to Kant; and Bonnet and Hemsterhuis' theories of perception (MPW 274–275).[12] The overall strategy, as it appears, was to play Hume against himself by appealing to his authority, yet at the same time drawing consequences from it that directly led to his own realism. This was as self-justifying realism, as Kant had claimed his own to be, but not, according to Jacobi, because of the grounds that Kant had adduced for his. In this, Jacobi was also playing Kant against himself.

Abstracted from its dialogue form, this is the argument.[13] Jacobi had been accused of irrationalism, even Papism, for having made knowledge rest on faith (*Glaube*) (MPW 262ff.). But Hume himself, the Berliners' darling,[14] held that the truths we subjectively hold most secure are in fact based on belief (also *Glaube*) (MPW 266ff.). There is, of course, an element of tendentiousness in this, for, as we noted earlier, "belief" does not carry in English the overtly religious connotations of the German "*Glaube*." Nonetheless, Jacobi could rightly claim that he was in accord

[12] Charles Bonnet (1720–1793), a widely known naturalist of the day; François Hemsterhuis (1721–1790), a popular elegant author.

[13] "Abstracted" and, accordingly, somewhat elaborated.

[14] Jacobi and Hamann associated the *Aufklärer*, the Enlightenment propagandists, with Berlin.

with Hume in holding that the attitude we assume in adhering to an evidence which is invincible yet does not admit of proof is appropriately called "belief." On this, Hume and Jacobi agreed. (Reid, incidentally, is also mentioned in context; MPW 266.) This is precisely the kind of evidence that Jacobi had called "revelatory" in his dispute with Mendelssohn, and had been criticized for it. Hume himself does not use the term, but, according to Jacobi, it is the term that common sense would use for the evidence delivered in sense-experience (*Sinnlichkeit*)—an evidence which, as Hume would grant, is based on the body's immediate feeling for reality (MPW 272, 278–279). Common language is testimony to this fact. All the terms denoting the mind's relation to reality are compounds of *Sinn*, a word which means, appropriately, both "sense" (as in "sensation") and "meaning." Jacobi delights in giving examples (MPW 303): *Unsinn* (nonsense), *Schwachsinn* (dullness of sense), *Stumpfsinn* (insensitivity), *Leichtsinn* (frivolity), *Scharfsinn* (sharpness of sense), *Tiefsinn* (profundity of sense), and *Wahnsinn* (madness or "being out of one's senses"). Moreover, it is the body, because of its immediate engagement with its environment, that restricts the imagination's power of embroidering the content of sense-experience at will if left unchecked. The body, in other words, makes possible the common distinction between fictitious and true belief.[15]

There is a problem, however. *Sinnlichkeit* is an affection of the body, a subjective state; consequently, so are also all the states of mind dependent on it. The question is whether the evidence that in sense-experience commands immediate assent is of reality *as affecting the body* or of reality *as it is in itself*. The issue here is not whether in experience we are in direct connection only with representations of reality internal to us rather than with reality in itself. This issue has already been settled. Sense-experience is revelatory as a matter of fact. The internal and the external coincide in it. The issue is rather whether the reality that thereby appears in the sequence of such experiences is held together in that sequence because of the experiences, the product of the experiencing body, or, on the contrary, it is the reality in itself that makes the sequence possible in the

[15] Jacobi invokes Hume's authority for this, citing his *Enquiry Concerning Human Understanding*, Section V, Part II (MPW 271).

first instance. At issue, in other words, is what holds reality together—whether the latter is a cohesive whole in itself or just a scattered aggregate of events only held together by a perceiver in imagination.

This is the problem which, traditionally, has been stated in terms of whether, in perception, one perceives a cause-effect relation connecting what is being perceived per se or whether the connection is instead added by the imagination. Hume, as Jacobi makes clear, famously denied that the connection is perceived directly. And Jacobi agreed with him in denying that, although not directly perceived, the connection's objectivity might nonetheless be established by ratiocinative inference. This is not possible, since any ratiocinative process would depend on representations (*Vorstellungen*), and these are only second-hand reproductions of what is given in *Sinnlichkeit.* Jacobi joined Hume in criticizing the empiricists and rationalists of the day (Mendelssohn included, in the case of Jacobi) for relying on inference for their alleged realism. Inasmuch as the issue at stake might be resolved at all, it would have to be done on the basis of *sense-experience* itself, because of its structure as direct sensing.[16] Hume, as Jacobi points out, had made a gesture in this direction by admitting that there is a feeling of power that pervades all our experiences, and that our idea of the cause/effect connection derives from it—"specifically, from the feeling of its use [namely, of the power] *in overcoming an obstacle*" (MPW 292). Nonetheless, this did not amount for Hume to the direct perception of a connection between presumed cause and alleged effect. And even when we can establish a reliably steady correlation between the two—as when we connect wanting to do something and actually seeing what we want done—the connecting itself is not perceived. It is established only *ex post facto*—in effect, by means of inference, and, therefore, with no direct evidential basis (MPW 291). In other words, according to Hume the fact that we have a generalized feeling of power is undeniable, but it is not for that reason the warrant for categorically asserting that there is, whether in us or in reality at large, any definably individual source of agency that would objectively link the experienced phenomena together (MPW 292–293).

[16] For Jacobi's rather disjointed argument, see especially MPW 278–279, 290–293.

Spinoza is also mentioned in context (MPW 292–293). Although his intellectual universe was completely different from Hume's, there was nonetheless a relevant parallel between the two that made the mention unavoidable. Both, while admitting causality in general—Spinoza, as a law of nature; Hume, as a generalized feeling of power—fell nonetheless short of admitting the reality of individualized agency. Both denied, in other words, the possibility of human freedom and human personality, the one issue closest to Jacobi's heart. In conversation with Lessing in 1780, Jacobi had not directly engaged Spinoza in argument. There was no common ground for the engagement, for Spinoza, as Jacobi believed, like all the rationalists of his kind, was given to walking on his head, ignoring immediate experience. Back then, therefore, faced by a Lessing who seemed comfortable in Spinoza's company, he had had no choice, as we have seen, but to appeal to his heart, that he redresses his position. Hume's case was however different. For Hume operated at the level of immediate experience, just like Jacobi. It was therefore possible for Jacobi to argue against him on that common ground—which is exactly what he proceeds to do in the Dialogue, in the process of turning Hume against himself (MPW 291–293).

Jacobi's objection was straight to the point. Hume had misrepresented the facts of experience by parsing them according to the language of speculative thought, in accordance with the latter's speculative interests. The result was that such concepts as that of the cause-effect relation cease to be recognizable in actual experience. As Jacobi had already said in conversation with Lessing, "the faculty of thought is always a spectator, and cannot ever be a source of action" (MPW 291). On that standpoint, therefore, the experiential content that gives meaning to the coming-to-be of something new, to "the *principium fiendi*, or *generationis*" (MPW 290), or, in other words, to the concept of causality as understood in common language, simply vanishes. The concept itself, Jacobi says, "would never have entered the language of beings who [, like the metaphysicians,] were *only capable of* [*intellectual*] *intuition and judgment*" (MPW 290). "But," Jacobi asks his interlocutor, preparing him for the decisive move in his argument, "is that the sort of beings that we are? *Surely*, my dear fellow, we can *also act*!" Indeed, *we act*. And Jacobi continues: "The non-speculative man was talking long before the philosophers began their discussion, and before some philosophers gradually managed to turn the use of language upside down. These philosophers wanted things to conform to words, just as the words previously had to conform to things" (MPW 290–292). And he adds, confirming Schelling's estimate that his historical manner of philosophizing had presaged his own:

> We know for instance that ancient peoples, or the uncivilized tribes of today, did not, or do not now have, such concepts of cause and effect as those that arose among more cultured peoples before or since. They see living beings everywhere, and they know of no power that is not self-determining. For them every cause is a living, self-manifesting, freely acting, personal power of this kind; and every effect is an *act*. And without the living experience of such a power in us, a power of which we are continuously conscious, which we use in so many ways, and which we can even let go of, without diminishing it—without this basic experience we should not have the slightest idea of cause and effect. (MPW 292)

All this amounted to saying that Hume, despite his commitment to the empirical method in philosophy, still was, in Jacobi's estimate, a rationalist.[17] He had not taken seriously in consideration what it is like, in lived experience, to be confronted by an obstacle—the very circumstance, as Hume himself had said, that precipitates our feeling of power. Neither would the obstacle be seriously felt *as obstacle* nor would we feel seriously constrained by it, unless, in one single and undivided event, the obstacle was recognized as deriving from a well-determinate source of agency, and we, for our part, would by the same token become aware of our own agency and the limits thereof (MPW 277). Jacobi was simply elaborating, for the edification of Hume, what he had already explained to Mendelssohn in dispute with him. There is no real agency unless there is a determinate source of it, and no awareness of it except in the encounter of one such source with another. There is no I without a Thou. The encounter, moreover, has to be originally at the level of the body. Although we might not be given in our philosophical culture to seeing living things everywhere,

[17] Jacobi does not say this explicitly, but the implication is obvious.

experience is nonetheless itself a life-form, and the evidence it affords a living one.

With this conclusion, Jacobi had disintricated himself from any ties he might hitherto have established with Hume by appealing to his authority. Hume ceases to figure in the Dialogue.[18] His one-dimensional view of experience has made his presence irrelevant. But, Jacobi asks, did that mean that he (Jacobi) had thereby secured a concept of causality, one "that pertains absolutely to the concept of the possibility of things in general?" It would seem, rather, that, when one derives a concept from experience, as Jacobi has just done, one "must forgo its absolute universality or necessity" (MPW 293). Jacobi himself raises the issue. Conceptually, of course, the issue is significant on its own merits. In context, however, its being raised also showed how much Kant had been in Jacobi's mind all along. This is the point in the argument where Jacobi turns Kant's transcendental idealism against itself.

4 Jacobi and Kant

Jacobi's argument had the markings of being transcendental in method from the beginning—with a difference, of course, as we shall see. Nonetheless, it is only at this point that its transcendental character appears deliberately staged. Jacobi begins by admitting that, if by saying that a concept is universal and necessary one means that it is self-validating precisely *as concept*, certainly he had not met this threshold in the case of causality. On the other hand, he continues, if one allows that it is possible to establish that to entertain such a concept is de facto necessary for "all finite beings endowed with reason," and that the concept "must lie at the base of *every* experience of theirs," it would follow that "the concept of cause and effect is a necessary one, a principle; and that the law of causal connection is a fundamental law governing the whole field of nature" (MPW 293). In other words, this is the line of argument, clearly

[18] I am following the logic of Jacobi's argument as it stands. However, there might have been other historical reasons for Hume disappearing at this point. As Jacobi advertises in the Prefatory Note, the present Dialogue was originally intended as three, only the first dedicated to David Hume (MPW 255).

transcendental, that Jacobi proposes. Granted a fact of experience, the point is to establish the conditions that make it possible: the circumstance that such conditions make the fact possible thereby establishes the necessity of the conditions themselves. In the present case, the fact at issue is the agency that we experience in action, as Jacobi has already established arguing against Hume. The point is to identify the conditions that make this acknowledged agency effectively such—that is to say, such as originates from a determinate source to which it can be attributed. And this is precisely the task that Jacobi proceeds to absolve in what follows—in the process, although not mentioning Kant by name, in fact reintroducing all his categories, that of causality included (MPW 293–296). Jacobi is broadening his argument against Hume by casting it in transcendental format.

The argument is long, yet, as one might object, too vague on details to be altogether convincing. Jacobi's is more the impromptu sketch of a phenomenology of experience than a fully elaborated theory. Nonetheless, the conclusion he reaches is not unwarranted. As he says,

> So, it would seem that we have shown the concepts of reality, substance or *individuality*, corporeal extension, succession, and cause and effect, to be concepts that must be common to all finite, self-revelatory, beings; and we have shown also that these concepts have their concept-independent object in *the things in themselves*—consequently they have a true, *objective*, meaning. (MPW 296)

He goes on:

> Concepts of this kind, however—the kind that must be given *in toto in every experience* and with such a degree of *primacy* that, unless *they* were objective, no concept would have an object [...]—such concepts have always in the past been called universal and necessary in an absolute sense; and the judgements and inferences derived from them have been called *cognitions a priori*. (MPW 296)

And finally, making the point he had been angling for all along:

> So we don't need to make *these fundamental concepts and judgments* into mere *pre-judgments of the understanding, whereby they become independent of experience.* [...] *As mere prejudices of the human understanding* they would be valid only for men and for the sensibility that is proper to humans; so they would be valid only under conditions which would, in my judgments, deprive them of all value. (MPW 296)

Jacobi is playing Kant against himself. The pages that follow are an implicit extensive criticism of transcendental idealism. Kant's aim had been to safeguard the possibility of making universal and therefore necessary judgements about the objects of experience, that is, to safeguard the possibility of intelligibility, without, however, abandoning the standpoint of experience itself. Jacobi's goal had been the same. But instead of displaying the required possibility in the structure of lived experience itself as Jacobi himself had just done, Kant, according to Jacobi, had instead assumed intelligibility to be a function of concepts per se independent of it. This is the factor that marked the difference between Jacobi's and Kant's argument. Kant had sought the source of the intelligibility in the reflective conditions of the understanding, an abstractive faculty that did not preclude the possibility that things are in themselves otherwise than its reflective conditions require. Nor, for that matter, was the possibility precluded on that procedure that such conditions did not apply to experience at all. The implication was that Kant had failed by the very standards he had set for himself.

5 Jacobi and Reason

So far as Jacobi was concerned, the source of intelligibility was first to be sought in the typically human body. To cite a medley of representative texts: "The perception of the actual and the feeling of truth, consciousness and life, are one and the same thing" (MPW 305). "Enter into yourself and search deeper there—search deeper and deeper—for what we call reason. You will find that you either hold the principle of reason to be one with the principle of life, or must make reason into a mere accident of a certain organized whole. As for me [Jacobi], I hold the principle of reason to be one with the principle of life" (MPW, 301). "And this really means that reason, as the distinguishing characteristic of man vis-a-vis animals, is only the characteristic of a particular sensibility" (MPW 302). "But where, since the time of Aristotle, is the philosophy to be found whose principles do not yield this same result? Where is there a philosophy that does not expound it in some form or other as a doctrine [...]? But then, more often than not, we let this reason, which has sprung from the senses, wondrously beget a miraculous youth, one which is supposedly equipped with his own special talents and powers for raising us far above the sphere of our sensations" (MPW 302–303). This was, of course, the reason of the *Aufklärer*. But, according to Jacobi, "*One never has more understanding than one has sense* [*Sinn*]" (MPW 303). "It follows that, with respect to all created beings, their rational cognition would have to be tested, ultimately, against their sensible one; the former must borrow its *validity* from the latter" (MPW 303).

Indeed, "Wie die Triebe, so der Sinn; und wie der Sinn, so die Triebe." Jacobi was reasserting the formula he had proclaimed to Mendelssohn. This was the part of Jacobi's Dialogue to which Schelling was especially alluding in his praise of Jacobi—with good reasons, because Jacobi had at least the sketch of a promising theory of experience that respected the latter's historical nature. The theory, nonetheless, still lacked an adequate internal self-limiting principle—the robust distinctions, as we have said, that would make the relation of *Sinn* and *Trieb* more than one of simple identity, as Jacobi himself certainly did not envisage it. This was a problem that could have been dealt with by further elaborating the given formula, and by further developing the theory of experience based on it which in the Dialogue was instead only adumbrated. Jacobi never did this.

This was also the part of the Dialogue that Jacobi later found embarrassing and, as Schelling rightly complained, obfuscated in the second edition, with added tendentious footnotes. The problem was that, while Jacobi's formula, on the face of it, conceptually resolved the heart/mind tension that affected the Enlightenment's culture, at some deep emotional level it apparently did not resolve it for Jacobi personally. His pietist side still had the last word. This is evident also in the Dialogue, where there is no lack of reminders that, pervading experience, there is the constant presence of the supernatural. As he says, "It is indeed a great advantage of our nature that we are capable of receiving from things the sort of impressions that display their manifold distinctly; hence of conceiving the inner word, *the concept*, for which we then create an outer being with a sound from our mouth and breathe the living soul into it" (MPW 306). Jacobi is here alluding to our sciences. He continues, "But these words begotten by finite seed are not like the words of *He Who Is*, and their life is not like the life of the spirit that calls being forth from nothingness. The moment we loose track of this infinite distinction, we remove ourselves from the source of all truth; we forsake God, Nature, and ourselves" (MPW 306). Perhaps. But how was Jacobi to retain this infinite distinction within the framework of a theory which tended rather towards the kind of naturalism that Schelling later advocated, and in which Spinoza and Goethe, in their different ways, would also have felt perfectly at home? This is a question that looms large in the Dialogue but is never openly confronted, let alone resolved. But it must have troubled Jacobi unconsciously even as he produced the work. The concluding part, not unlike the conclusion of the *Spinoza-Letters*, is but a prolix evocation of intimations of the supernatural. It is as if Jacobi were making amends for the naturalism he had just flirted with.

Much happened after the *Spinoza-Letters* and the *David Hume*. We know, however, how, around 1800, Jacobi finally resolved the heart/mind tension. He did it exactly how Schelling complained. On the one hand, he attributed to reason the intuitive power to apprehend being that the rationalist philosophers also attributed to it. He, however, characterized it as feeling, as if he could thereby absolve it of its abstractness. On the other hand, by thus sublimating feeling as reason—by removing it from nature, in other words—he also removed from the latter the element where, on his earlier testimony, one would recognize the creative power of God's word. To cite: "Reason does not produce concepts, it builds no system, and does not even judge: instead, *like the external senses*, it simply reveals, it makes positive proclamations" (MPW 562). "And so we admit without fear that our philosophy begins with feeling, but with a feeling that is *objective* and *pure*; that it professes the authority of this feeling to be the highest; and that, in its role as a doctrine of the supersensible, it bases itself on this authority alone" (MPW, 363–364). Hamann's fear had come true. Jacobi had become a pillar of the Church, and the Church was exactly as Schelling thought—still that of the Enlightenment.

References

Beiser, Frederick C. 1987. *The Fate of Reason: German Philosophy from Kant to Fichte*. Cambridge, Mass.: Harvard University Press.
di Giovanni, George. 1994 (augmented paperback edition 2009). *The Unfinished Philosophy of Friedrich Heinrich Jacobi*. In *The Main Philosophical Writings and the Novel Allwill*, Friedrich Heinrich Jacobi, translated and edited by George di Giovanni, 1–167. Montreal & Kingston: McGill-Queen's University Press.
———. 2005. *Freedom and Religion in Kant and his Immediate Successors: The Vocation of Humankind, 1774–1800*. Cambridge: Cambridge University Press.
———. 2021. *Hegel and the Challenge of Spinoza: A Study in German Idealism, 1801–1831*. Cambridge: Cambridge University Press.
Goethe, Johann Wolfgang. 1774. *Götter, Helden und Wieland*. Leipzig: Auf Subscription.
———. 1777/1787. *Triumph der Empfindsamkeit*. Leipzig: Göschen.
Hamann, Johann Georg. 1955–1979. *Briefwechsel*. Vols. I–VII. Edited by W. Ziesemer, and A. Henkel. Wiesbaden: Insel Verlag.
Jacobi, Friedrich Heinrich. 1994 (augmented paperback edition 2009). *The Main Philosophical Writings and the Novel Allwill*. Translated and edited by George di Giovanni. Montreal & Kingston: McGill-Queen's University Press.
Kant, Immanuel. 1999. *Correspondence*. Translated and edited by Arnold Zweig. Cambridge: Cambridge University Press.
Schelling, F.W.J. 1994. *Schelling on the History of Modern Philosophy. Translated by Andrew Bowie*. Cambridge: Cambridge University Press.

7

Kant on Driving Forces: Parallels and Differences in Kant's Conceptualization of *Trieb* and *Triebfeder*

Manja Kisner

1 Introduction

In his philosophical works, Kant uses the concepts of *Triebfeder* and *Trieb*, but the meaning of *Triebfeder* is much more clearly elaborated and discussed than the definition of *Trieb*. The concept of *Triebfeder* is important for Kant's practical philosophy and appears in all his major moral writings as well as his lectures on ethics. Although there are some particular queries concerning the use of this term in Kant, it is uncontested that *Triebfeder* plays an integral philosophical role in those works and thus belongs to his essential philosophical terminology.

The concept of *Trieb*, in contrast, is much less clearly defined, and its meaning remains in many ways ambivalent, although the English translation of this term—as drive—is less problematic than of *Triebfeder*. In the

M. Kisner (✉)
University of Wuppertal, Wuppertal, Germany
e-mail: kisner@uni-wuppertal.de

M. Kisner, J. Noller (eds.), *The Concept of Drive in Classical German Philosophy*,
https://doi.org/10.1007/978-3-030-84160-7_7

second part of his third *Critique*, Kant refers to Blumenbach's concept of *Bildungstrieb* (formative drive) and, accordingly, uses this term in a biological sense, although there are specific differences between Blumenbach's and Kant's understanding of *Trieb*.[1] In addition to this meaning, Kant uses this concept in other, more figurative senses, for instance, when he speaks about *Trieb zur Wissenschaft* (drive to science) or *Trieb zur Erweiterung unserer Erkenntnis* (drive to expand knowledge). In this regard, it is not obvious whether this concept indeed has a particular philosophical meaning and if it is really an integral part of Kant's philosophical apparatus.

When it comes to English translations of *Trieb* and *Triebfeder*, further problems arise. In Kant's scholarship, *Triebfeder* is commonly translated as "incentive." Nevertheless, as Timmermann (2007, 180) and Herrera (2000, 395) point out, the English word "incentive" is misleading in this context because it hides the mechanical origin of *Triebfeder*. The word "incentive" describes in English an (external) object that motivates someone to action. However, Kant's *Triebfeder*, in its most frequent use and thereby following the original mechanical meaning of *Triebfeder*, referred to an inner motivation for action and not an external object. Therefore, Herrera (2000, 395) suggests that the most appropriate English translation of Kant's concept of *Triebfeder* is a "driving mechanism," although in Kant, of course, this driving mechanism must be understood figuratively, describing the driving mechanisms of our actions. Timmermann (2007, 180), furthermore, describes *Triebfeder* as "a motivating desire, the force that propels an agent forward if he or she so chooses."

Moreover, in some English translations, the difference between *Triebfeder* and *Trieb* disappears. Grenberg (2001), for instance, translates *Triebfeder* simply as "drive," paving the way for her interpretation of Kant and emphasizing the important role of feelings (also sensible feelings) in Kant's theory of rational action. In this way, she favors a particular interpretation of Kant, which might—among other things—also result from her decision to interpret *Triebfeder* as drive, that is, *Trieb*.

The aim of my paper is to clarify the ambiguity connected with Kant's depiction of *Trieb* and *Triebfeder* in his works. Although Kant himself

---

[1] For a discussion of Blumenbach's influence on Kant, see the text by Zammito in this volume.

does not discuss the difference between these terms, I believe that we can explain these differences philosophically if we focus on the underlying imagery that accompanies each use. According to my interpretation, Kant's theory of action, in which his concept of *Triebfeder* plays a crucial role, is still constructed on the model of the functioning of machines. In its original mechanical meaning, *Triebfeder* referenced the inner springs in the machines, for instance, inside a clock; this picture, I argue, influenced and served as a model for Kant's account of practical *Triebfeder*.

The concept of *Trieb*, on the contrary, is based on another model, that of organisms, and hence points to a more organic and dynamic account of inner motivations and inner driving forces. In this case, the focus is on a process—Kant uses drives to depict the process of self-organization and self-formation. Although the philosophical meaning of *Trieb* can be, according to my account, most clearly reconstructed on the basis of Kant's example of organisms and inner purposiveness in the second part of his third *Critique*, its conceptual novelty extends beyond this particular biological context. In what follows, I argue that Kant applies this idea to move toward a new conception of a system as a purposive, self-organizing, and dynamic process, which will later prove essential to the German idealists. The philosophical apparatus connected with *Trieb* is therefore very different from that of *Triebfeder*.

2 *Trieb* and *Triebfeder*: Etymology

The concepts of *Trieb* and *Triebfeder* were introduced into German philosophy prior to Kant. In the early Enlightenment thought of the first half of the eighteenth century, Thomasius and Crusius discussed the concept of *Trieb* in their moral and psychological writings, whereas for Baumgarten, the concept of *Triebfeder* was of particular importance.[2] The question about the influence of these and other sources (such as the physiological and biological body of research from the eighteenth century) on Kant's understandings of *Trieb* and *Triebfeder* is very complex and exceeds

[2] For a helpful overview on the use of the concept of drive in pre-Kantian German philosophy, see Buchenau (2002). For further discussion of Thomasius, see Heidemann in this volume.

the scope of this chapter. The main aim of this brief historical reflection is to call attention to subtle differences in the meaning of these two terms prior to Kant, which proves essential to any review of the differences in his notions of *Trieb* and *Triebfeder*.

First, Kant uses the concept of *Triebfeder* as a German translation of the Latin term *elater animi* (animal spring; CPrR, 5:72).[3] However, the word "elater" was unknown to classical Latin, and the exact origins of the concept of *Triebfeder* remain unclear.[4] Despite these etymological difficulties, in the eighteenth century, the concept of *Triebfeder* first had a mechanical meaning, designating the spring inside a machine, as the Latin word "elater" confirms. In accordance with this understanding, Grimms' dictionary (1854–1961) defines *Triebfeder* as a "driving steel spring (*treibende stahlfeder*)" within a machine, for instance, a clock.[5] Hence, *Uhrfeder* (the spring of the clock) describes the *Triebfeder* of the clock. Similarly, in Adelung's dictionary (1811), *Triebfeder* is termed "an elastic spring (*eine elastische Feder*), insofar as it causes the parts of a machine to move," and the clock springs are called *Triebfedern*. According to this meaning, the *Triebfeder* could be appropriately translated into English as the driving mechanism of a machine. This use is consistent with an increased interest in mechanical terminology beginning in the seventeenth century but also indicates the impact of machine metaphors on the broader, figurative use of *Triebfeder*, which became more and more important during the eighteenth century.

Taken in the latter extended sense, *Triebfedern* were used to describe bodily functions and the movements of animals as well as, more concretely, the driving forces that propel agents to action. The *Triebfeder* can designate the springs of the soul (*Triebfedern der Seele*) as well as the springs of an action (*Triebfedern der Handlung*). However, in a figurative sense, it is not always clear if *Triebfeder* indeed references inner driving forces or if it rather refers to an object or idea that prompts the agent to act. This second understanding is suggested in Adelung's dictionary,

---

[3] References to Kant's texts are made by citing the volume and the page number of the *Akademie-Ausgabe* (Kant 1900ff.). Translations of Kant are drawn from *The Cambridge Edition of the Works of Immanuel Kant* (cf. Kant 1999, 2000 in the reference list).

[4] See Schweiger (1999, 161f.), and Sala (2004, 161f.).

[5] Translations from Grimms' and Adelung's dictionaries are my own.

when *Triebfeder* is described figuratively as "any idea, any thing, which determines the acting force (*wirkende Kraft*) in us to activity (*Tätigkeit*)." In this sense, the *Triebfeder* can also be conceived as a motivating ground (*Bewegungsgrund*) and matches Baumgarten's definition of *Triebfeder* as a *Bewegungsgrund* in his *Metaphysica*, where the representations function as driving causes of the soul (see Buchenau 2002).[6]

In contrast to the mechanical origins of the concept of *Triebfeder*, the word *Trieb*, which was used in spoken language as early as the thirteenth century and became widespread in literary language beginning in the sixteenth century, has a much broader meaning. Adelung's and Grimms' dictionaries note the origins of this term in its verbal use—as *treiben*—describing the process of being driven to something.[7] In line with this meaning, Adelung emphasizes that *Trieb* can be understood as "an abstraction of the verb *treiben*" (*das Abstractum des Zeitwortes treiben*). With the noun *Trieb*, hence, the young shoots of a tree or a plant, which recently sprouted, can be described. Moreover, with the word *Trieb*, the act of being driven (*die Handlung des Treibens*) is described, for instance, the act of driving cattle to pasture, and most importantly for later meanings, *Trieb* also references the driving force of a living being (*die Kraft eines lebendingen Geschöpfs*), which can also be called an instinct or a natural drive (*Naturtrieb*).

In this latter sense, *Trieb* is relevant for describing the functioning of nature and illustrates the strivings and tendencies of organisms, for example, "as pushing and promoting from within" (*als das drängen, das fördern von innen heraus*; Grimm 1854–1961). This meaning is reflected also in Thomasius' notion of *Trieb* as an inclination or a tendency, which he used to describe human emotions and decision making; he speaks about "the drive of the will" (*der Trieb des Willens*, Grimm 1854–1961; see also Buchenau 2002). This meaning corresponds with Grimms' definition of

[6] We can nonetheless ask if the identification of *Triebfeder* and *Bewegungsgrund* is completely appropriate. Timmermann, for instance, suggests that in Kant's *Groundwork* there is a difference between *Triebfeder* and *Bewegungsgrund*. The latter Kant uses to describe the motivating ground, namely the "object that prompts the mechanism of motivation to action" (Timmermann 2007, 181). In this sense, the *Bewegungsgrund* is an object and describes something static. This difference between *Triebfeder* and *Bewegungsgrund* in Kant will be discussed further in Sect. 4.

[7] Here, I just point to the few most relevant meanings of *Trieb* for my analysis.

Trieb as *Antrieb* (motor, incentive), in which the effect is not directed toward a specific object but rather describes the process of striving.

This brief comparison of the original uses of *Trieb* and *Triebfeder* confirms that the concept of *Trieb* had a much broader meaning than *Triebfeder*[8] and that the concept of *Triebfeder* originates from the more general term *Trieb*. This is also confirmed by the further use of *Trieb* for something that drives something else into motion, for example, the driving wheel that drives another thing (*Treibrad* or *Triebrad*; Adelung 1811). This use of *Trieb* matches the mechanical meaning of *Triebfeder*. Thus far, both terms can be used to describe the activities of machines, animals, even humans.

Despite these common origins, however, we also must point to subtle but important differences in the use of these two terms. The concept of *Trieb* seems to emphasize the striving character more and, as such, describes a tendency or inclination toward something. This is also the meaning that we find in Thomasius' use of *Trieb* when discussing a psychosomatic tendency, which is in humans irreducible to reason and represents the activity of the will. The concept of *Triebfeder*, conversely, does not emphasize as much the activity or the dynamic process as such, but rather focuses on pointing out the driving cause behind the activity. However, as I pointed out earlier, the figurative use of *Triebfeder* remains ambivalent if the cause is, as in this case, the representation of the object or rather the inner force that propels actions. In any case, Kant will try to resolve this ambivalence in his practical philosophy by distinguishing more clearly between *Triebfeder* and *Bewegungsgrund*, as we will see later. Thus far, we can conclude that the concept of *Trieb* is less focused on a particular object as a driving cause of action but rather emphasizes the dynamic activity of the process as such.

[8] This can be confirmed with a quick glance at Grimms' dictionary, in which the entry *Trieb* has over twenty pages, whereas the entry *Triebfeder* takes up only two pages.

3 Kant on Machines: The Example of the Clock in the Third *Critique*

In his works, Kant refers to both notions, *Triebfeder* as well as *Trieb*, although he dedicates a much more substantial and explicit analysis to the former. Therefore, I will start with the discussion of Kant's notion of *Triebfeder*, as he uses it in his practical philosophy, before proceeding to an analysis of *Trieb* in the third *Critique*. Before I turn to these two topics, I will first briefly sketch Kant's account of machines in the third *Critique* because it is here that we can infer how he describes mechanical *Triebfedern* in philosophical terms. To show this, I will focus on §65 of the second part of the third *Critique*, in which he expounds on the difference between machines and organisms. Although he does not mention the concept of *Triebfeder* explicitly here, his clock example gives us important clues on the matter. The following passage neatly summarizes the main characteristics of machines, in this case a clock:

> In a watch one part is the instrument for the motion of another, but one wheel is not the efficient cause [*die wirkende Ursache*] for the production of the other: one part is certainly present for the sake of the other but not because of it. Hence the producing cause [*die hervorbringende Ursache*] of the watch and its form is not contained in the nature (of this matter), but outside of it, in a being that can act in accordance with an idea of a whole that is possible through its causality. Thus one wheel in the watch does not produce the other, and even less does one watch produce another, using for that purpose other matter (organizing it); hence it also cannot by itself replace parts that have been taken from it, or make good defects in its original construction by the addition of other parts, or somehow repair itself when it has fallen into disorder: all of which, by contrast, we can expect from organized nature. (CJ, 5:374)

We can infer from this passage that for the characterization of machines, the difference between efficient and producing causes is crucial. Kant claims that a wheel cannot function as an efficient cause for the production of another wheel. This does not mean, however, that according to Kant's view, wheels are not efficient causes. Of course they are, although not for production, but only for the motion of other wheels. The producing cause of the clock must lie outside of it, namely in the being that imagines and constructs a clock. This distinction between two kinds of causes corresponds, as I argue, with Kant's twofold account of causality, that is, with the difference between efficient (*nexus effectivus*) and purposive causality (*nexus finalis*), which Kant discussed in the passages preceding the clock example.

When one wheel (or other parts of the clock, for instance, the spring) causes the motion of another wheel, we must describe this connection in terms of efficient causality.[9] This connection Kant uses to refer to all kinds of connections and motions of empirical objects, which follow mechanical laws. However, when we want to define the producing cause of the clock as a whole, we need to reach out for purposive causality. Kant calls purposive causality the causality "in accordance with a concept of reason (of ends)" or "the causality of a concept with regard to its object" (CJ, 5:220), that is, conceptual causality.[10] In this case, the producing cause of the clock is the clockmaker, or more precisely, the idea or the concept of the clock that the clockmaker has and tries to realize when constructing it. As a result, it is due to the producing cause of the clock that we conceive of the clock as a machine and not as a naturally appearing object.

What Kant tries to emphasize here is that, in order to grasp the specific features of all human-made artifacts, efficient causality does not suffice, and the reference to purposive causality is needed because it is this causality that explains the object as a result of the idea and conceives it as a purposive object. Insofar, the existence of the clock is dependent upon the external agent (as an external producing cause) who imagines and constructs the clock, and, accordingly, the clock has a special function, for example, it performs a specific task of measuring the time.

But how can Kant's account of efficient and purposive causality help us to expand on the concept of *Triebfeder*? The mechanical *Triebfedern* literally describe the elastic springs of the clock, and as such, these springs have exactly the same causal role as all other parts of the clock machinery:

[9] Kant defines efficient causality in the third *Critique* as descending, meaning that the thing which functions as an effect cannot at the same time be the cause and, hence, the series must be conceived as descending (see CJ, 5:372).

[10] For further discussion of the importance of conceptual causality for Kant, see Teufel (2011).

they function as efficient causes, which bring to motion other parts of the clock. As *Triebfedern*, however, they are also different from other efficient causes in nature because they are dependent upon the particular function of the clock and are hence in service of the producing causes and the corresponding purposive causality. In this sense, they are constitutive of the correct or incorrect functioning of the clock; if one spring breaks, for instance, the whole clock will stop measuring the time correctly. Therefore, we refer to the mechanical concept of *Triebfeder* in particular when we discuss machines and their motive powers (*bewegende Kräfte*; CJ, 5:374) but not when we refer to physical forces of nature. Even though these motive powers obey the laws of nature and act in accordance with efficient causality, they are at the same time in service of fulfilling the purpose of the clock as a machine. In sum, the mechanical concept of *Triebfeder* describes the constitutive parts of the driving mechanisms of the machine and is, as such, dependent upon the function that the machine has to fulfill.

4 Kant's Practical Notion of *Triebfeder*

Drawing on the results of the previous section, I now turn to Kant's notion of *Triebfeder* in his practical philosophy. For this purpose, I will focus on the three main works of Kant's ethics, the *Groundwork of the Metaphysics of Morals*, the second *Critique*, and the late *Metaphysics of Morals*. In all of these texts, Kant mentions the concept of *Triebfeder*, although there are some differences in how he defines this term. Most importantly, I will argue that also, in these cases, the notion of purposive causality can help us to further expand on Kant's account of practical *Triebfedern*.

Kant's notion of *Triebfeder*, as he uses it in practical philosophy, describes either empirical or moral *Triebfedern*, that is, the *Triebfedern* of pure reason. In *Groundwork*, Kant does not yet mention the *Triebfedern* of pure reason; instead, he defines the effects of our actions "as ends (*Zwecke*) and incentives (*Triebfedern*) of the will" (G, 4:400). Such actions, driven by *Triebfedern*, cannot have an unconditional moral worth. The concept of *Triebfeder*, as understood here by Kant, is a posteriori and material, distinguishing it from the formal principle of the will, which is a priori. This description seems to suggest that *Triebfedern* are the external objects (or representations of these objects) which motivate our actions and, as such, function as empirical *Triebfedern*.

In another passage in his *Groundwork*, Kant then further expands on *Triebfeder* by distinguishing it from the concept of *Bewegungsgrund*. As we have seen in Sect. 2, historical dictionaries sometimes conflate these two notions, but in his *Groundwork*, Kant defines them in different ways. *Bewegungsgrund* is, according to Kant, an objective ground of the will, whereas *Triebfeder* is only the subjective ground of desire. This concurs with the description from the previous passage: *Triebfedern* are subjective and material, meaning that they are a result of particular ends expressed by agents. Objective ends depend on *Bewegungsgründe*, which hold for every rational being; therefore, the *Bewegungsgrund* can in this sense only be the principle of the will as such, namely the moral law. Subjective ends, conversely, are relative in determining the will empirically. In this understanding, *Triebfedern* are identified as empirical *Triebfedern*, and therefore, the conception of a *Triebfeder* of pure reason, which becomes important in the second *Critique*, would be, as Allen Wood claims, a contradictory term from the perspective of *Groundwork*.[11]

However, Kant does not stick to this definition of *Triebfeder* consistently through all his works. Already in *Groundwork*, Kant also offers another non-empirical definition of *Triebfeder* when he describes the respect for the law as a *Triebfeder* of actions with moral worth (G, 4:440).[12] This meaning then comes close to how he defines *Triebfedern* in the sec-

---

[11] Therefore, Allen Wood concludes that the *Triebfedern* of pure practical reason, which are the focus of the second *Critique*, should be from the perspective of *Groundwork* conceived as *contradictio in adjecto*: "Thus a ground having objective validity (valid a priori for all rational beings) cannot count as *Triebfedern* in this sense, even if the will is in fact determined by it (i.e., even if an agent chooses to act on it), and the term '*Triebfedern* of pure practical reason,' with *Triebfedern* taken in this sense, would be a *contradictio in adjecto*. Hence there is a need in the *Groundwork* for a contrasting term (*Bewegungsgrund*), referring to grounds of the will's possible determination which do have this objective validity" (Wood 1999, 361).

[12] The discussion about the role of moral feelings and the feeling of respect in Kant's account of *Triebfedern* must be left aside in this limited analysis. I also do not go into the details of the antagonistic interpretations of intellectualists and affectivists, who hold different positions when it comes to the question of whether moral feeling can be understood as a motivational force. For the most recent and comprehensive overview of this topic and of different standpoints, see Kassan (2019, 82–86).

ond *Critique*, in which an entire chapter—*Von den Triebfedern der reinen praktischen Vernunft*—is dedicated to the *Triebfedern* of pure practical reason. In this chapter, Kant mentions the Latin term *elater animi*, that is, an animal spring, as an equivalent of the German word *Triebfeder* and further expands on it with the help of the difference between objective and subjective determining grounds of the will.

Although the distinction between objective and subjective determining grounds of the will from the second *Critique* sounds similar to the distinction between the objective ground of the will and the subjective ground of desire, there is a crucial difference in how Kant defines the two terms in each work. Allen Wood notes that, after *Groundwork*, hence also in the second *Critique*, the subjective ground "is taken to mean any determining ground [...] that determines the will, including grounds that have objective validity, when these are considered as possibly determining the will" (Wood 1999, 361). As a result, in the second *Critique*, the moral law can function as an objective as well as a subjective determining ground of the will.

This difference has important consequences for Kant's notion of *Triebfeder* in the second *Critique*. According to his altered account, the moral law functions as an objective determining ground when it determines the will immediately; only in this case actions have moral worth (CPrR, 5:71). However, because in the case of finite rational beings, the moral law does not determine the will necessarily, the same moral law must at the same time also function as the subjective determining ground.[13] In this case, the moral law is conceived as the *Triebfeder* of moral actions. As we see here, in addition to the empirical *Triebfedern*, we can also speak of the *Triebfedern* of pure practical reason. This means that we can distinguish between moral (when the moral law itself is the *Triebfeder*) as well as non-moral *Triebfedern* (when inclinations or self-love function as a *Triebfeder*).[14]

[13] To clarify this, he points out the difference between the holy will, which cannot have any *Triebfedern*, and the human will, in which we can distinguish between subjective and objective determining grounds (CPrR, 5:72).

[14] The difference between moral and non-moral *Triebfedern* corresponds to the difference between the autonomy and the heteronomy of the will.

This understanding of *Triebfedern* points to an important shift of focus between *Groundwork* and the second *Critique*. In the latter work, *Triebfeder* is not used to describe only sensible *Triebfedern* that determine our will, but instead portrays the inner driving forces of the will, which can be either moral or non-moral. This meaning resembles the original mechanical meaning of *Triebfeder*, which is used to describe the inner motive powers of machines and, as such, illustrates their driving mechanism. In accordance with this interpretation, Stephen Engstrom argues that Kant's concept of *Triebfeder* almost always—meaning when we focus on the second *Critique* and not on the few disparate passages contained in *Groundwork*—describes "something in the subject that generates the action, rather than an object or circumstance that prompts it" (Engstrom 2010, 92). Moreover, he defines Kantian *Triebfedern* as "sources of willing and action, not their ends, outcomes, or effects" (Engstrom 2010, 92). This concurs with my interpretation, which claims that Kant's *Triebfedern* can be conceived—analogous with the mechanical *Triebfedern*—as the inner driving springs of actions, which can be either empirical or moral.[15]

In the case of moral laws, this means that the objective determining ground of the will tells us how we should act, and the *Triebfeder* as the subjective ground describes our inner motivation, upon which we act. Jeanine Grenberg therefore maintains that *Triebfedern* play a crucial role in "bringing about the existence of a particular object or state of affairs" (Grenberg 2001, 155).[16] This reading is further refined in Kant's *Metaphysic of Morals*, where he distinguishes between the moral law playing a legislative role and hence being objectively necessary, and the *Triebfeder*, which plays an executive role and describes "the actual deter-

[15] Cf. Engstrom (2010, 93): "So far as the moral law is a subjective determining ground of the will, it is a spring. It is a cognitive representation, but one that has force, or efficacy."

[16] I agree with this particular interpretation of *Triebfeder* as causing the agent to act, although Grenberg also has a much stronger and more controversial thesis that "for Kant, all action of finite rational agents involves being impelled [*angetrieben*] to action by drives [*Triebfedern*], that is, by the sensible influence of feeling on the faculty of desire" (Grenberg 2001, 155). Grenberg interprets *Triebfeder* as a "drive" in order to emphasize the sensible influence that *Triebfedern* have on the faculty of desire (Grenberg 2001, 155). This identification of *Triebfedern* with sensible drives is, according to my understanding, not mandatory, and it concerns a particular controversy between affectivists (to which Grenberg belongs) and intellectualist interpretations, which I cannot discuss in greater detail here.

mination of a being to action, that is, choice," as Grenberg describes it (2001, 156). In the second case, the focus is not on the objectively valid law but on the determination of the will as *Willkür* to a specific action.

This short summary of Kant's views on *Triebfeder* allows me to draw some important parallels with the clock example in the third *Critique* and to look again at the relevance of Kant's notion of purposive causality. Kant refers to purposive causality not only when he characterizes the specifics of artifacts and machines but also in the context of his theory of action. In both contexts, our faculty of desire operates in accordance with concepts and is identified as purposive causality. As Kant points out, everything that is represented through a will is determined to causality through concepts (CJ, 5:172). This conceptual causality is crucial for describing technical products as well as actions.

However, there is also a crucial difference between the use of purposive causality in a moral context and in all other non-moral contexts. In the first case, it is the concept of freedom that determines the will, and Kant speaks in this regard about morally practical principles. In the second case, it is the concept of nature that determines the will, and the resulting principles are technically practical principles. As a result, Kant's notion of purposive causality is just as relevant for describing machines and other artifacts as it is for describing actions that happen out of respect for the moral law. The only difference is that, in the latter case, we have to deal with "practical legislation of reason in accordance with the concept of freedom" (CJ, 5:171), whereas in other instances, we are concerned with the legislation of reason in accordance with the concepts of nature. Moral law, which is given to pure practical reason, is, accordingly, conceived as "a law of causality through freedom" (CPrR, 5:47).

What relevance does purposive causality have for Kant's practical account of *Triebfedern*? I will try to expand on this by way of comparison between mechanical *Triebfedern* and empirical and moral *Triebfedern* in Kant's practical philosophy. In all cases, I argue, *Triebfeder* functions as a relational concept dependent upon a concept that determines purposive causality. As we have seen previously, mechanical *Triebfedern* are the inner driving mechanisms of machines and hence play a crucial role in realizing what we could call the purpose of the machine. The concept of the machine as such, which functions as a producing cause in Kant's imagery and is related to purposive causality, determines the tasks of the inner parts of the machine, for instance, elastic springs as *Triebfedern*, which then guarantee the proper functioning of the machine.

With regard to the figurative notion of *Triebfeder*, Kant offers, as we have seen, a somewhat different definition of this term, but generally, we can distinguish in his account between empirical and moral *Triebfedern*. In both cases, *Triebfeder* can be viewed as a relational concept. First, we can interpret Kant's notion of an empirical *Triebfeder* in *Groundwork* as a relational concept because, on the one hand, he emphasizes its relationship to particular purposes or objects that can determine our will—purposes or objects can again be understood as concepts that determine purposive causality—and on the other hand, he also points to our sensible inner inclinations. Although Kant is, in this example, not completely clear as to whether he wants to understand *Triebfeder* as an external object that motivates our actions or as an inner driving force, I think that the most persuasive interpretation, also with regard to his later works, is to recognize that the reference to empirical *Triebfedern* is meaningful only when both aspects are present, namely some external ideas or objects that influence our desires as well as our inner inclinations that make us responsive to those external purposes. Empirical *Triebfedern* are thus related to purposive causality, and the concept to which this conceptual causality relates is, similar to mechanical *Triebfedern*, the concept of nature.

Second, in the case of moral *Triebfedern*—I here refer to the definition in the second *Critique*—we can apply a similar schema. On the one hand, the moral law as an objective determining ground of the will functions as a concept that determines purposive causality. The moral law provides us with the rules according to which we should act; in this sense, practical reason has the ability to determine our will according to the concept, that is, the moral law. In contrast to the first case, however, moral law as a concept is here not a concept of nature (for instance, some external object), but a concept of freedom, and therefore, here we have to engage in purposive causality through freedom.

However, in humans and other finite rational beings, concepts and ideas—either in the form of moral or technical principles—do not have the executive powers to determine the will necessarily. The final realization of ideas is due to other factors. Here, again, the similarity between

Triebfedern as the inner motivational forces of our actions and as the inner mechanical drives of machines comes to the fore. *Triebfedern* guarantee that a machine will function properly or that we will act according to the moral law. Therefore, the moral law has to function as a subjective determining ground of the will and as a *Triebfeder*; in this sense, *Triebfeder* describes the inner driving force that prompts agents to act just as the mechanical *Triebfeder* guarantees the proper functioning of the clock.

In sum, despite the differences in those cases, Kant's conception of moral *Triebfedern* can be envisaged according to the same schema that we have used previously for mechanical *Triebfedern*. Most importantly, this comparison confirms that the notion of *Triebfeder* is used in the contexts in which efficient causality does not suffice, but where instead, a reference to purposive causality is needed. As a result, the *Triebfedern* of either machines or human actions play an important role in the realization of purposive causality: they do not describe how we or the machines should act, but they instead describe the inner motivations upon which we or they do act.

5 The Concept of *Trieb* in Kant's Third *Critique*

Zooming out from Kant's account of *Triebfeder* in his practical philosophy, I will now turn to the analysis of Kant's concept of *Trieb* in the third *Critique*. *Trieb* is not one of the key concepts of Kant's philosophy, and in the second part of the third *Critique*, it seems to play a marginal role. Kant refers explicitly to the notion of *Bildungstrieb*, which he adopts from Blumenbach, in one passage of the *Methodology of the Teleological Power of Judgment* (CJ, 5:424), in order to distinguish it from mechanical forces. Moreover, another related concept is that of a formative power (*bildende Kraft* or *Bildungskraft*), which comes to the fore on few occasions in the second part of the third *Critique* (see CJ, 5:374, 419, 423, 424). In the first part of the third *Critique*, as well as in other works by Kant, the concept of *Trieb* is also mentioned here and there, but its use in these contexts remains conceptually vague and mostly metaphorical.[17] My focus here will therefore be only on the concept of *Trieb* as it is understood in the second part of the third *Critique* because it is in this context that we can outline the philosophical meaning of *Trieb* most clearly.

In contrast to the seemingly marginal role of *Trieb* in Kant's philosophy, I argue that its relevance is much greater than we might expect at first glance, especially when we discuss it in view of Kant's notion of internal purposiveness, a very important but also highly controversial principle that Kant introduces in the second part of the third *Critique* in order to describe the unique features of organic nature. In this way, an important shift of focus from purposive causality to internal purposiveness and from *Triebfeder* to *Trieb*, respectively, appears.

I will first turn my attention to §65, which was already discussed in Sect. 3, but which also offers a starting point from which we can grasp the difference between organisms and human-made machines. In this context, Kant defines organisms as self-organizing beings to which we ascribe special formative powers (*bildende Kräfte*) instead of motive powers (*bewegende Kräfte*) characteristic of technical products such as the clock:

> An organized being is thus not a mere machine, for that has only a motive power, while the organized being possesses in itself a formative power, and indeed one that it communicates to the matter, which does not have it (it organizes the latter): thus it has a self-propagating formative power, which cannot be explained through the capacity for movement alone (that is, mechanism). (CJ, 5:374)

In this quote, we first see that motive powers are used to explain mechanical movements and, accordingly, the functioning of machines. The *Triebfedern* of the machines, such as the springs or the wheels of the clock, have motive powers. Organisms, in contrast, cannot be conceived as mere machines, although they share some similarities; most

[17] For a discussion of the concept of drive in the first part of the third *Critique* and Kant's political writings, see Heidemann in this volume.

importantly, we consider both kinds of objects to be purposive, in contrast to other objects in nature, such as stones, for instance.[18]

Nonetheless, organisms differ crucially from machines because of their self-propagating and self-organizing abilities, which Kant tries to explain further by presupposing the special formative powers of organisms. This notion of formative powers is, insofar, congruent with Kant's notion of *Bildungstrieb*, as he defines it in §81 in order to point out the organizational characteristics of organisms that cannot be confined to merely mechanical forces (CJ, 5:424), and I will hence use these two terms interchangeably.

How can we philosophically define and explain these formative powers or drives? Although Kant remains vague on this point, I will try to show that Kant gives us in the third *Critique* at least an implicit answer to this question. Let me discuss the difference between motive and formative powers a bit more in detail. As we have seen previously, according to Kant's understanding, machines are dependent upon a concept as a cause of the object; in the case of the clock, the clockmaker, or to be more precise, their idea of the clock functions as the producing cause of the clock. Therefore, efficient mechanical causality does not suffice to describe machines as technical products; for this purpose, a reference to the purposive causality of the producer is needed. Motive powers within the machine hence depend upon the function that a specific technical product has; this means that their task is to fulfill the aims of the purposive causality.

Kant's account of purposive causality points to the shortcomings of this procedure when we try to grasp organisms as self-organizing beings. In this case, we cannot presuppose an external, producing cause of an

[18] Organisms share with machines a reciprocal relationship between the parts and the whole; for both kinds of objects, it holds that the parts stand in relation to the whole. However, there is a difference between organisms and machines with regard to this relationship: "But if a thing, as a natural product, is nevertheless to contain in itself and its internal possibility a relation to ends, i.e., is to be possible only as a natural end and without the causality of the concepts of a rational being outside of it, then it is required, second, that its parts be combined into a whole by being reciprocally the cause and effect of their form. For in this way alone is it possible in turn for the idea of the whole conversely (reciprocally) to determine the form and combination of all the parts: not as a cause—for then it would be a product of art—but as a ground for the cognition of the systematic unity of the form and the combination of all of the manifold that is contained in the given material for someone who judges it" (CJ, 5:373).

organism, and hence the reference to purposive causality does not help us to explain the specific, self-organizing features essential to organisms. Organisms have to be described "without the causality of the concepts of a rational being outside of it" (CJ, 5:373), which we presuppose in the case of artifacts and machines. This is only possible when the parts that are combined into a whole are "reciprocally the cause and effect of their form" (CJ, 5:373), meaning "that it is required that its parts reciprocally produce each other" (CJ, 5:373) and, furthermore, "produce a whole out of their own causality" (CJ, 5:373). Therefore, Kant gives us the following definition of an organism: "An organized product of nature is that in which everything is an end and reciprocally a means as well" (CJ, 5:376).

This account of organisms leads Kant to a special internal causality, according to which an organism is itself a producing and efficient cause. Because in this case, the external producing cause is missing, this internal causality is not real purposive causality; instead, it is only a principle of internal purposiveness that we can postulate. The solution that Kant offers in this context is, as we can see, to introduce a new principle, namely the principle of internal purposiveness, in order to explain the self-producing and self-organizing feature of organisms that purposive causality is not able to explain properly.

The meaning of this principle, however, is far from clear, and Kant is well aware of the problems that this notion brings about.[19] According to Kant, we only recognize two different kinds of causalities, efficient and purposive causality, which play a constitutive role in his theoretical and practical philosophy, respectively. The principle of internal purposiveness, in contrast, is only a regulative principle that a reflecting judgment presupposes in order to judge objects that we cannot explain sufficiently by either efficient or purposive causality. As such, the principle of internal purposiveness cannot be equated with purposive causality, although Kant introduces this principle in the third *Critique* through "a remote analogy with our own causality in accordance with ends," that is, our purposive causality (CJ, 5:375).

That Kant encounters difficulties in explaining this term, we also see in the way he introduces it. Kant is not able to philosophically deduce this

[19] For detailed discussions on Kant's notion of internal purposiveness see Zuckert (2007).

principle but instead introduces it by means of examples, or, as he claims, by deriving it from experience (CJ, 5:376)—for instance, with the help of the tree example through which he identifies the specific features of organisms.[20] The strategy that Kant applies here is to postulate the principle of internal purposiveness heuristically because this enables us to grasp and explain the difference between other natural objects and organisms as living beings. Therefore, Kant calls this principle "a maxim for the judging of the inner purposiveness of organized beings" (CJ, 5:376).

However, Kant rejects the possibility of turning the principle of internal purposiveness into a real internal causality, comparable to either efficient or purposive causality. What he does instead is to set up a connection between internal purposiveness and formative powers or drives. Both terms are closely related and support each other: they are used in order to point out the self-producing and self-organizing features of organisms. Moreover, with these two terms, Kant enters uncharted territory, which calls for novel philosophical terminology that will be able to describe dynamic and self-organizing processes in a more appropriate way.

On this basis, we can develop, I suggest, the following definition of Kant's concept of *Trieb*: this concept is used for describing processes for which we cannot assume an external creator as a producing cause, but which we, in contrast, conceive as self-sufficient and self-producing processes. In this case, these processes cannot be explained with the help of producing causes and correlated motive powers; instead, we must assume that special inner formative powers cause these dynamic purposive processes.

This leads to a new conception of living processes. Machines, as technical processes, reference objects with a function that remains the same throughout the process—they do not evolve and develop; even though they follow an internal dynamic, this dynamic does not change during the process.[21] The purposes and the functions of the machines are predetermined and dependent upon the idea of the purpose and the creator of

[20] These specific features of organisms—the ability of self-generation, self-formation, and reproduction—are all due to the self-organizing abilities of organisms. See CJ, 5:371.

[21] This holds at least for simple machines, such as the clock, which Kant has in mind here, but I do not claim that these machines are the only possible examples. For self-generating machines, Kant's conceptual apparatus related to his views about technical causality would not suffice.

the machine. Organisms, as living processes, describe, in contrast, processes that develop and progress. We cannot recognize the external producing cause of those objects, and therefore, the living processes remain unfinished. We cannot ascribe to them a final purpose; this is only possible when we can assume a creator as a producing cause of the object, but this external cause is missing in this case.

The concept of drive as a formative power is used by Kant in order to describe the striving and evolving character of living processes. As a result, the concept of drive describes a tendency or inclination to do something, but the direction of this tendency is in these cases not set up by a rational agent, and therefore, it does not have the same meaning as a purpose, which is determined by its creator. The terminology that we need to describe organisms is hence different than in the case of machines: whereas the characteristic features of machines are described through purposive causality and motive powers, the organizing features of living processes are described by drives and internal purposiveness.

6 A Comparison of Kant's Practical *Triebfeder* and His Concept of *Trieb*

To conclude, I will now return to the comparison between the notions of *Triebfeder* and *Trieb* that initiated my discussion, but now I can discuss these differences from the Kantian perspective. Although Kant does not always use these two terms in a clear and unambiguous way, we can nonetheless point out some of the relevant differences between them. Although they have similar origins, the *Triebfeder* brings about, due to its mechanical terminology, a different contextual background than the *Trieb*. This is something that we can observe quite clearly in Kant's understanding of these two terms. Kant's moral philosophy and his theory of action rely, according to my analysis, on a very technical account of how we act, which still resembles our understanding of the functioning of machines.

When he tries to distinguish between moral and non-moral actions, the main question Kant asks relates to the determining ground of the will; this is the question about the cause of our actions. Similarly to how the clockmaker is the (producing) cause of the clock, we as rational agents are the cause of our actions, meaning that we determine our will. Accordingly, either the moral law or some other technical-practical principle determines our will. Therefore, we need purposive causality to describe our actions. The *Triebfeder* is in this illustration conceived as a force that then brings about the realization of the law. The imagery that Kant gives us here is a static one; he does not describe the strivings and temptations of the will or the conflicts and tensions of the will. *Triebfeder* describes the inner motor of the will, but this power is not perceived as a formative and self-producing power. Here, the background model is not that of organic matter but rather that of the machines.

However, when Kant refers to the concept of *Trieb*, the underlying picture shifts. With *Trieb*, he describes a living process, which has to reconcile different tensions and is, as such, much more dynamic and in constant development. The accompanying terminology therefore has to change as well; here, it is the concept of internal purposiveness that calls for special formative powers in order to describe these processes as self-organizing and self-producing processes. The conceptual and terminological background for the philosophical understanding of *Trieb* is developed by Kant in the second part of the third *Critique* when he discusses the special status of organic nature.

However, I do not want to claim that the concept of *Trieb* needs to be limited only to this biological context. On the contrary, I think that what Kant indicates in the second part of the third *Critique* is the need for a new conception of dynamic processes. It is for this reason that Kant uses *Trieb* in non-biological contexts, for instance, when he speaks about the *Trieb zur Erweiterung unserer Erkenntnis* or about the *Trieb zur Wissenschaft*. What is common to all these different uses of *Trieb* is that Kant uses this concept to identify a special kind of formative and self-organizing processes, which cannot be grasped through purposive causality and motive powers alone.

References

Adelung, Johann Christoph. 1811. *Grammatisch-kritisches Wörterbuch der hochdeutschen Mundart*. Vol. 4. Wien: Bauer. https://lexika.digitale-sammlungen.de/adelung/online/angebot.

Buchenau, Stefanie. 2002. Trieb, Antrieb, Triebfeder dans la philosophie morale prékantienne. *Revue germanique internationale* 18 (*Trieb: tendance, instinct, pulsion*): 1–24.

Engstrom, Stephen. 2010. The *Triebfeder* of Pure Practical Reason. In *Kant's Critique of Practical Reason. A Critical Guide*, ed. Andrews Reath and Jens Timmermann, 90–118. Cambridge: Cambridge University Press.

Grenberg, Jeanine M. 2001. Feeling, Desire and Interest in Kant's Theory of Action. *Kant-Studien* 92: 153–179.

Grimm, Jacob, and Wilhelm Grimm. 1854–1961. *Deutsches Wörterbuch*. http://woerterbuchnetz.de/cgi-bin/WBNetz/wbgui_py?sigle=DWB.

Herrera, Larry. 2000. Kant on the Moral *Triebfeder*. *Kant-Studien* 91: 395–410.

Kant, Immanuel. 1900 ff. *Gesammelte Werke*. Edited by the Royal Prussian (later German) Academy of Sciences. Berlin: De Gruyter.

———. 1999. *Practical Philosophy*. Edited and translated by M. J. Gregor. Cambridge: Cambridge University Press.

———. 2000. *Critique of the Power of Judgment*. Translated by P. Guyer, and E. Matthews. Cambridge: Cambridge University Press.

Kassan, Nora. 2019. *Reine praktische Vernunft fühlen. Kants Theorie der Achtung*. Berlin/Boston: De Gruyter.

Sala, Giovanni B. 2004. *Kants "Kritik der praktischen Vernunft". Ein Kommentar*. Darmstadt: Wissenschaftliche Buchgesellschaft.

Schweiger, Clemens. 1999. *Kategorische und andere Imperative. Zur Entwicklung von Kants praktischer Philosophie bis 1785*. Stuttgart-Bad Cannstatt: frommann-holzboog.

Teufel, Thomas. 2011. Kant's Non-Teleological Conception of Purposiveness. *Kant-Studien* 102: 232–252.

Timmermann, Jens. 2007. *Kant's Groundwork of the Metaphysics of Morals. A Commentary*. Cambridge: Cambridge University Press.

Wood, Allen W. 1999. *Kant's Ethical Thought*. Cambridge: Cambridge University Press.

Zuckert, Rachel. 2007. *Kant on Beauty and Biology: An Interpretation of the Critique of Judgment*. Cambridge: Cambridge University Press.

8

The Drive to Society in Kant's *Critique of the Power of Judgment*

Dietmar Heidemann

1 Introduction

On the face of it, the connection between drive and teleology in Kant's third *Critique* seems fairly natural. For drive can be conceived as a natural property pertaining to living beings in the sense that living beings are caused by nature to strive, develop or evolve relative to an end, in order to achieve an aim or satisfy a purpose. Being determined by nature in view of an end or purpose reflects that "drive" is a relational concept because whatever is driven is driven towards an end, purpose, or in a certain direction. However, apart from the casual remarks on Blumenbach's "formative drive" ("Bildungstrieb", CJ, 5:424) as the living being's power to organize itself, and nature's "drives" as guiding "fetters" (CJ, 5:432),

D. Heidemann (✉)
University of Luxembourg, Esch-sur-Alzette, Luxembourg
e-mail: dietmar.heidemann@uni.lu

M. Kisner, J. Noller (eds.), *The Concept of Drive in Classical German Philosophy*,
https://doi.org/10.1007/978-3-030-84160-7_8

there is no mention at all of "drive" in the Teleology of the third *Critique*.[1] One might therefore suspect that, in the third *Critique*, a *teleological* use of "drive" is absent. Things look quite different in the Aesthetics. Here Kant refers to "drive" specifically as "the drive to society", presumably "natural to human beings" (CJ, 5:295), and similarly to "the vigorous drive towards the lawful sociability" (CJ, 5:355). Contrary to what one might expect, Kant's use of "drive" in the *Critique of the Power of Judgment* is predominantly *practical* rather than *teleological*, that is, it concerns issues of moral and political philosophy rather than philosophy of biology. This might come as a surprise, although there are obvious reasons for why this is so. One important reason is that Kant stands in a tradition that understands "drive" as a natural property almost exclusively within the domain of the social rather than within the field of the biological. The contexts of the Aesthetics where Kant employs the term clearly indicate this: The human being's empirical interest in the beautiful depends on the social, and, as we shall see, for this reason also on the cultural condition of human development. Kant conceives of the social or cultural condition of the beautiful, or more precisely of "beautiful art", in terms of the connection of "the universal feeling of participation" and universal communicability (CJ, 5:355). It is "the vigorous drive towards the lawful sociability" (CJ, 5:355) that makes this connection possible.

From the exclusively *practical* use of "drive" in the Aesthetics and its absence in the rest of the third *Critique* it does not follow, though, that Kant ignores the systematic function of "drive" in the Teleology. In what follows, I argue that although Kant does not mention "drive" in the Teleology, the practical or social meaning of "drive", as already alluded to in the Aesthetics, is nonetheless pivotal in the Teleology's doctrine of the ultimate end of nature. More precisely, the systematic connection between *drive* and teleology becomes apparent in the philosophy of history and culture as presented by Kant in the "Methodology of the Teleological Power of Judgment". It is here where he further develops his earlier views, especially those of the *Idea of a Universal History with a Cosmopolitan*

[1] References to the *Critique of the Power of Judgment* (CJ) are from Kant (2000). For all works from Kant the standard citation format is: volume number of the Akademie-Ausgabe (Kant 1900 ff.) and page number.

Aim, that nature, by means of the drive to society, promotes the historical development of mankind and culture as its ultimate end. I proceed in three steps: In Sect. 1, I sketch the historical background of Kant's use of "drive" in early modern philosophy. This background explains why Kant, in the third *Critique* and other writings, sticks to the social meaning of "drive". In Sect. 2, I look into how Kant implements the social drive in the first systematic account of the connection between history and teleology, that is, the *Idea*. In Sect. 3, I discuss the systematic connection that Kant establishes between drive and teleology in the third *Critique*. I argue that Kant proposes a conception of "drive" that integrates the natural and cultural dimension of human life. In the conclusion I address the limits of this conception.

2 The Drive to Society: Historical Background

In the *Religion Within the Boundaries of Mere Reason* (1793), Kant distinguishes three classes of "original predisposition to good in human nature" (*Religion*, 6:26–28): (1) the "predisposition to *animality*" as *physical* self-love, (2) the "predisposition to *humanity*" as *rational* self-love, and (3) the "predisposition to *personality*" as moral respect. Kant specifies the "predisposition to *animality*" as non-rational self-love and brings the latter under three headings: (1) (the drive to) "self-preservation", (2) "propagation" through the "sex drive", and (3) "community with other human beings" through the "drive to society" ("Trieb zur Gesellschaft") (*Religion*, 6:26). In the *Religion* Kant conceives of the "drive to society" like in the third *Critique* (CJ, 5:295) not just as physical but explicitly as "mechanical self-love", that is, as a kind of "love" "for which reason is not required" (*Religion*, 6:26). The idea of the *physical* or *mechanical* "drive to society" is in line with what he already explicated some six years earlier than the publication of the third *Critique*. In the fourth proposition of the *Idea for a Universal History with a Cosmopolitan Aim* (1784), he declares that "to enter into society" is a "propensity" or "predisposition [...] in human nature" that "nature employs" in order to foster the social and cultural development of mankind (*Idea*, 8:20). That reason is not required for the drive to society and that the "inclination to become socialized" belongs to human's "natural predispositions" (*Idea*, 8:20–21)[2] does not turn the social drive into an irrational drive. *Physical* or *mechanical* rather indicates that the drive to society cannot be reduced to rational capacities or powers. This is clear from the historical background of the Kantian use of "drive to society".

The Kantian conception of the "drive to society" is deeply influenced by the seventeenth- and eighteenth-century debate of "drive" in moral and psychology textbooks by authors such as Thomasius, Rüdiger, Wolff, Baumgarten and Crusius.[3] The debate rotates around the question of whether or not the drive to society is reducible to human's rational powers, that is, whether it can be classified as an independent *natural* drive. Here I can only sketch the debate along general lines. The context is the discussion of human nature, directed against Descartes, and the status of rational capacities. In that respect Christian Thomasius seems to be of particular importance. In his *Ausübung der Sittenlehre* from 1696 he insists, unlike Descartes, that thought is not what makes up human nature: "It is hence equally false that human nature, by means of which he is distinguished from non-rational animals, is nothing over and above thought" (Thomasius 1696, 81; my translation).[4] Thomasius values "the propensity and drive of the will to be a much more distinguished power of the human souls compared to the thinking of the understanding" (Thomasius 1696, 81; my translation).[5] The crucial point is that for

[2] In the *Anthropology*, by contrast, Kant depicts human socialization as rationally motivated: "The human being is destined by his reason to live in a society with human beings and in it to cultivate himself, to civilize himself, and to moralize himself" (7:324 f.). Contrary to this, he says in the same work that humans "feel destined by nature to [develop], through mutual compulsion under laws that come from themselves, into a cosmopolitan society (cosmopolitismus) that is constantly threatened by disunion but generally progresses toward a coalition" (7:332; my emphasis).

[3] For a very instructive and more detailed account of this background see Buchenau (2002). Buchenau's historical reconstructions are very helpful for understanding the genesis of the concept "drive to society" in early modern philosophy. I do not, however, agree with how she classifies Kant in that history (see footnote 11).

[4] The German original reads: "Und also ist es ebenmäßig falsch, daß das Wesen des Menschen, wodurch er von den unvernünfftigen Thieren unterschieden wird, eintzig und alleine in Gedancken bestehe."

[5] "Denn die Neigung und der Trieb des Willens ist eine viel edlere Krafft der Menschlichen Seelen als das Dencken des Verstandes."

Thomasius the "passion" of the will is "inclination", and the "operation" of the will is "choice" conceived as the will's "drive" (Thomasius 1696, 79).[6] This view is again opposed to Descartes. For Thomasius all operations of the soul depend on the passions since no one can think or reflect something without having sensed it before, and no one can make a choice independently of feeling an inclination.[7] The will is, therefore, the "power" ("Krafft") of the soul that "drives" the human being to act. Consequently, Thomasius does not conceive the will's choice as the product of an entirely rational operation of the mind but as the outcome of natural inclination and drive.[8]

Thomasius' general view that drive is a non-rational, natural property impacts his more specific conception of the drive to society. Accordingly, the social drive, too, is non-rational and therefore a natural propensity. In his *Von der Kunst Vernünffftig und Tugendhafft zu lieben* (1692) he states that the human being aims to promote other human's wellbeing, not only because each individual sees his or her wellbeing dependent on the common good but more specifically because the human being realizes that "God has willed it so and therefore gave him the drive to desire to live in other human beings more than in himself" (Thomasius 1692, 148).[9] This idea of the drive to society as a natural, physical drive implanted by God in the human being has been highly influential in early modern philosophy. Not only Thomasius' students like Rüdiger but also Wolff and Baumgarten have elaborated on the idea of drive (to society) as a non-rational property of human nature.[10]

---

[6] "Die Leidenschafft des Willens heißt Neigung, die Thäthigkeit, Wahl und Willkür, oder der willkührliche Trieb."

[7] Cf. Thomasius (1696, 79): "Ohne die vorhergehenden Leidenschafften sind die Thäthigkeiten der Seelen nichts würckliches. Denn wer kan etwas nachdencken, das er nicht zuvorhero sinnlicher Weise empfunden; und wer kan etwas erwehlen, wenn er gar keine Neigung dazu bey sich gespühret."

[8] Cf. Thomasius (1696, 81): "So ist demnach der Wille eine Krafft der Menschlichen Seelen, vermöge welcher der Mensch zu etwas geneiget wird, und hernach sich selbsten antreibet etwas zu thun oder zu lassen." Cf. also Buchenau (2002, 12–14).

[9] Cf. Thomasius (1692, 148): "Er suchet hiernächst anderer Menschen neben sich ihr wohl seyn zu befördern, nicht so wohl, weil von dem allgemeinen wohl seyn auch sein eigenes dependiret, sondern weil er erkennet, daß es GOtt so haben wolle und ihm deshalben einen Trieb gegeben, daß er in andern Menschen mehr als in sich selbst zu leben verlanget."

[10] For further details and references cf. Buchenau (2002, 16–20).

To provide another speaking example that is historically closer to Kant, Crusius, in his *Anweisung vernünftig zu leben* (1744), elaborates what one might term a *philosophy of drive*. He distinguishes several fundamental drives ("Grundtriebe", Crusius 1744, § 111) as fundamental powers. One is the drive to perfection ("Vervollkommnungstrieb", Crusius 1744, § 111), the other one is the drive to society (Crusius 1744, § 120). The drive to society stands in connection with the natural moral drive to love other human beings (Crusius 1744, §§ 125, 130). Crusius defines the drive to society as "the desire to unify with such objects in which we perceive perfection" (Crusius 1744, § 120). For Crusius the drive to society is independent and cannot be derived from the drive to self-perfection. Similar to Thomasius it is a natural drive and as such irreducible. As we will see in the following section, Kant basically continues this line of argument, that is, he conceives of the drive to society in terms of an independent drive that nature employs in order to promote the historical and cultural development of the human being. It is the teleology of the third *Critique* that provides the theoretical framework for his conception of the drive to society.[11]

3 The Drive to Society in the *Idea for a Universal History*

The systematic connection between the drive to society and teleology in the *Critique of the Power of Judgment* builds on the philosophy of history. Much of what the third *Critique* expounds on the function of the social drive within the alleged purposeful development of natural predispositions in the course of history can be traced back to the *Idea for a Universal History with a Cosmopolitan Aim*. In this section, I shall therefore discuss the role of the drive to society in the *Idea* before I turn to its function in the third *Critique*. Although, in comparison to the *Idea*, Kant does not change his principal views about the role of the social drive in the third

[11] I think Buchenau (2002, 24) is right that in his moral philosophy Kant breaks with the seventeenth-/eighteenth-century anthropological foundation of morals like in Thomasius. But unlike Buchenau I think that Kant in fact continues the *naturalist* conception of the drive to society of that age (see Sects. 2 and 3).

Critique, he substantially expands the impact it has on domains other than history.

In the *Idea*, Kant articulates his overall argument for the teleological consideration of history in three main steps. In the first three of the nine propositions, he states his main thesis that "[a]ll natural predispositions of a creature develop themselves completely and purposively" in accordance with a presupposed "teleological doctrine of nature" (*Idea*, 8:18). Kant specifies, though, that "[i]n the human being (as the only rational creature on earth), those predispositions whose goal is the use of his reason were to develop completely only in the species, but not in the individual" (*Idea*, 8:18). This complete development, compliant with the aim of the natural dispositions, is determined as a natural effort accomplished by the human being in the pursuit of practical goals.[12] Kant justifies his completeness-claim partly on the basis of observation of natural processes. The argument, however, is ultimately analytic because on his view it is contradictory to believe in teleology of nature while, for example, organs do not achieve their natural purposes. The teleology of nature and subsequently of history is hence presupposed and as such not deduced.[13] Under that premise, it would in fact be contradictory to hold teleology but deny—in principle—the purposeful development of nature and history. Propositions four to six answer the question of how nature manages to cause natural dispositions in the human species to develop completely and in conformity with an aim. According to Kant, the "means" nature employs in order to achieve this goal is the "antagonism in society, insofar as the latter is in the end the cause of their lawful order" (*Idea*, 8:20). The "inclination" aroused by the conflicts between individuals to socialize is the drive to society to which nature appeals.[14] Kant writes:

[12] Cf. *Idea*, 8:19.

[13] In that respect I disagree with Allison (2009, 34) who understands the completeness-claim as a "corollary" of Kant's overall teleological principle.

[14] Cf. *Idea*, 8:20 f. Although Kant's famous "unsociable sociability of human beings" (Idea, 8:20) is an oxymoron, it nonetheless is an expression of the drive to society as a natural inclination or propensity. Schneewind (2009) provides a very helpful account, especially of the historical context of this oxymoron.

> Thanks be to nature, therefore, for the incompatibility, for the spiteful competitive vanity, for the insatiable desire to possess or even to dominate! For without them all the excellent natural predispositions in humanity would eternally slumber undeveloped. The human being wills concord; but nature knows better what is good for his species: it wills discord. (*Idea*, 8:21)

Kant considers the arrival into civil society under legal conditions (of freedom) as "the greatest problem for the human species" (*Idea*, 8:21). He sees in the institution and the guarantee of the civil constitution the true foundation of the development of natural dispositions, in such a way that he goes on to assert that "a perfectly just civil constitution, must be the supreme problem of nature for the human species" (*Idea*, 8:22). According to the Kantian conception, civil society will, however, only be reached at a later stage within the historical development of the human species. This later stage is achieved when the legal community is subject to the law of freedom. As Kant finally explains in propositions seven to nine of the *Idea*: "The problem of establishing a perfect civil constitution is dependent on the problem of a lawful external relation between states and cannot be solved without the latter" (*Idea*, 8:24). The "antagonism" that exists at the level of individuals is repeated at the level of the existing states, that is, at the level of international interstate relations. It is supposed to be overcome by a "united power" (*Idea*, 8:24) of states within a federalist framework. This, too, is the work of nature that *drives* societies and states to federalize:

> Nature has therefore once again used the incompatibility of human beings, even of great societies and state bodies of this kind of creature as a means to seek out in their unavoidable antagonism a condition of tranquility and safety […], namely, to go beyond a lawless condition of savages and enter into a federation of nations. (*Idea*, 8:24)

For Kant, international federative structures count as the condition of development of the natural dispositions of the human being. Since this development is accomplished in the history of the human species, it is not restricted to politics but includes human "culture", that is the development of "art", "science" and "morality" (*Idea*, 8:24). Kant draws the following conclusion:

> One can regard the history of the human species in the large as the completion of a hidden plan of nature to bring about an inwardly and, to this end, also an externally perfect state constitution, as the only condition in which it can fully develop all its predispositions in humanity. (*Idea*, 8:27)

The situation thus reached would be at the same time the "universal cosmopolitan situation" (*Idea*, 8:28). Like civil society, the cosmopolitan situation originates in the natural drive to society as the "propensity to enter into society" (*Idea*, 8:20). Kant thus considers it to be possible, at least theoretically, "to work out universal history according to a plan of nature that aims at the perfect civil union of the human species" (*Idea*, 8:29). The function he assigns to the drive to society in the *Idea* has its primary application in the field of civil constitution and international relations. In the third *Critique*, Kant widens its scope by integrating the historical development of culture and morality to a much greater extent.

4 Drive to Society and Ultimate End in Kant's Third *Critique*

The teleological interpretation of history that Kant brings to the point in the ninth proposition of the *Idea* by way of concluding that one can "attempt to work out universal world history according to a plan of nature" (*Idea*, 8:27) is a claim that remains largely ambiguous. For is this conclusion cognition from *theoretical* reason, or should it rather be conceived as *teleological* cognition, that is, cognition merely in accordance with the principle of purposiveness that the reflective power of judgement refers to nature as such? Although one should not expect historical processes to be the object of reflective consideration on nature of the power of judgement, it is in fact the *Critique of the Teleological Power of Judgment* that provides an answer to this question. In this context Kant implements, once again, the idea of the drive to society as natural propensity of the human being to not only enter into the civil constitution but to even move forward to the federalization of states.

For Kant, the concept of purpose is pivotal not only for comprehending organisms but also in the context of the philosophy of history. In the third *Critique* he defines "purpose" as "the concept of an object insofar as it at the same time contains the ground of the reality of this object", such that "purposiveness" must be understood as "the correspondence of a thing with that constitution of things that is possible only in accordance with ends" (CJ, 5:180). The concept of "*purposiveness*", or more precisely: the "*purposiveness of nature* in its multiplicity" as "the principle of the power of judgment" (CJ, 5:180), must be differentiated as follows: *first*, as universal purposiveness of nature, *secondly*, as subjective *purposiveness* in Aesthetics and, *thirdly*, as logical-objective purposiveness in teleology. While universal purposiveness as the principle of the reflective faculty of judging directs us to conceive of nature in its diversity as a unity structured according to the classificatory, hierarchical order of universal and particular concepts, principles or laws, aesthetic purposiveness designates the harmonious play of imagination and understanding in the intuitive consideration of forms. By contrast, objective purposiveness represents organisms and the causal reciprocity of their parts as real purposes of nature.[15]

More specifically, when reflecting on nature, the reflective power of judgement presumes that there is, among the particular (empirical) laws in nature, a taxonomy that is arranged according to species and kinds. It presupposes the merely contingent fact that "the order of nature in its particular laws, although its multiplicity and diversity at least possibly surpass all our power of comprehension, is yet fitted to it" (CJ, 5:187). Now, the understanding seeks to discover this "order of nature" with the aim to "introduce into it unity of principles", that is to say to determine this "unity" according to an end that the power of judgement attributes to nature "because the understanding cannot prescribe to it any law on this matter" (CJ, 5:187). In cases where this "intention" or "end" is achieved, for instance by way of discovering the "unifiability of two or

[15] On Kant's division of the kinds of purposiveness and its application to history see Allison (2009, 29–31).

more empirically heterogeneous laws of nature under a principle that comprehends them both" (CJ, 5:187), the feeling of pleasure ensues.[16]

The principle of universal purposiveness of nature is valid for both, the subjective and the objective case, in which purposiveness can be represented in an empirically given object. This is possible either *subjectively* because of formal "agreement", independent of concepts, or *objectively* because of "a correspondence of its form with the possibility of the thing itself, in accordance with a concept of it which precedes and contains the ground of this form" (CJ, 5:192). Therefore, the universal purposiveness of nature for the reflective power of judgement makes possible both the subjective-aesthetic and the objective-logical purposiveness. Since philosophy of history is concerned with judging causal connections between actions and, in particular, with the question of an *organic* wholeness of history, the principle of aesthetic purposiveness is not applied in this field. For in the case of history the principle, which aims at the harmony of imagination and understanding in the intuitive consideration of forms, is otiose. According to Kant, judging history is possible only by means of the principle of objective purposiveness, according to which organisms must be represented in the reciprocal causation of their parts as real ends of nature. For history cannot be understood as a mere "aggregate of human actions" (*Idea*, 8:29) that constitute it. It rather must be conceived *organically*, that is as a purposeful whole that must first be presupposed before being filled by its parts, that is, the actions.

Kant develops this view further in the "Methodology of the Faculty of Teleological Judgment" (CJ, §§ 81–83). It is here where he resumes his earlier conception of the *Idea* that nature uses the antagonism between human beings and hence the drive to society in order to establish, in the historical development of the human species, "*civil society*" and "a cosmopolitan whole, i.e., a system of all states" (CJ, 5:432), such that the natural dispositions of humankind can develop completely and purposefully. In the third *Critique* Kant's argument is more elaborate than in the *Idea*, not only because it belongs to the context of the theory of purposiveness as such but also because Kant justifies the establishment of civil society and international federative structures specifically in connection with the

[16] Cf. Heidemann (2010).

doctrine of the ultimate end of nature. The argument can be reconstructed as follows: In § 81, Kant, first, specifies the relation between mechanism and teleological explanation of nature and then continues, in § 82, with a line of thought that leads from the "teleological system in external relations of organized beings" to the federative system of states as the purpose of history. The starting point is the idea that external purposiveness is added to inner purposiveness. External purposiveness means that "one thing in nature serves another as the means to an end", such that the question can be asked "why does it exist?" (CJ, 5:425). Two answers are possible to this question: Either the thing does *not* exist due to a cause that acts intentionally, such that it is simply a product of natural mechanism, or there is an "intentional ground of its existence" (CJ, 5:426). The latter is the case for organized beings, whose internal possibility itself is based on intentional causality. That is to say, the existence itself of a thing, to which external purposiveness pertains, must be conceived as purpose. This can be conceived in two ways: Either the "purpose of existence" of an organized being resides in that being itself, in which case it is not just an end, but at the same time a "final end" (*letzter Zweck*), or the end "is outside of it in another natural being", such that the thing exists according to an end, but not as the "ultimate end" (CJ, 5:426).[17]

Kant thus develops a system of natural ends whose focal point is an ultimate end. This ultimate end cannot be discovered *in* nature, though, since in principle everything in nature can always be thought of as a means for other purposes. The ultimate end must therefore lie *outside* of nature, in the sense that Kant puts it as a metaphysical vanishing point at the top of the system of natural purposes. Nevertheless, there is a "final end" (*letzter Zweck*) in nature which is "the ultimate end of the creation here on earth" (CJ, 5:426). This ultimate end is, according to Kant, the human being itself (CJ, 5:426). Here, Kant does not incorporate theological reflection but argues that the human being is the only being on earth who has "understanding", "because he is the only being on earth who forms a concept of ends for himself and who by means of his reason

[17] Allison (2009, 40) is exactly right that "[a]lthough a final end (in the sense of the highest goal or purpose) is also an ultimate end, the converse need not hold". For a clear exposition of Kant's complicated reasoning see Düsing (1968, 206–237).

can make a system of ends out of an aggregate of purposively formed things" (CJ, 5:427). The human being is therefore the ultimate end of nature because he or she possesses the capacity to set ends and to choose the appropriate means for achieving them.

The principle of external purposiveness is the linchpin of the argumentation upon which the subsequent foundation of civil society and of international federative structures is based. For reason applies this principle also to the end-means relation with respect to organisms in order to once again think of "a system of all the kingdoms of nature in accordance with final causes" (CJ, 5:427). What Kant calls here "system of all the kingdoms of nature according to causes", then figures as "the system of all states" (CJ, 5:432).[18]

The human being is thus not just like all other organized beings (organisms) an end of nature but at once the ultimate end of nature on earth "in relation to which all other natural things constitute a system of ends" (CJ, 5:429). Conceiving of the human being as the ultimate end of nature is, however, not to be misinterpreted as objective cognition since it is the reflective and not the determining power of judgement that raises this claim. This limitation also applies to the status of the hypothesis of purposive history. The purposiveness of history can only be thought in accordance with the principle of reflective judgement and should hence not be misunderstood as positive, objective cognition of the conformity of history to a plan. At this crucial point of the argument, Kant makes the transition towards the concept of culture and by implication to the concept of drive to society. The fact that the human being must be considered as the ultimate end of nature indicates that nature in general must be useful to humankind as an end such that it is to be conceived as a means for the human being. We have, Kant argues, once more two options: *Either* nature satisfies this end in such a way that the end consists in the achievement of "happiness" as promoted by nature *or* the human being, as the ultimate end, appropriates nature in such a way that "it is the aptitude and skill for all sorts of ends". It is in this second case that the ultimate end of nature consists in the "culture" of the human being

[18] For details of the teleological fundament of federalism and cosmopolitanism in Kant see Heidemann (2016, 98–110).

(CJ, 5:430). And as we already know from the *Idea*, culture is a manifestation of the drive to society and results from natural antagonism among human beings.[19]

Already in the *Groundwork of the Metaphysic of Morals*, Kant argues that "happiness" cannot be "the proper end of nature" (4:395), not least because the concept of happiness is ambiguous and essentially subject-dependent. Similarly, in the third *Critique*, he considers "happiness" to be an inappropriate *end* because happiness is what the human being cannot attain anyway, "for his nature is not of the sort to call a halt anywhere in possession and enjoyment and to be satisfied" (CJ, 5:430). Would happiness of the human being be in fact the ultimate end of nature, then the human being itself would be nothing other than a means. For in this case the ends the human being sets himself or herself would ultimately be themselves means relative to the achievement of happiness. Also, since happiness represents an "ultimate end" that is supposed to be determined as to its content but that is inaccessible for us, it cannot be what the human being, according to Kant, shall in general pursue, namely an "ultimate end" (*Endzweck*) outside of nature (CJ, 5:431). Granted this argument, one might nevertheless wonder whether the human being does not yet function as a means to an end. For through the drive to society nature causes the human being to build communities and therefore to develop culture as the ultimate end of nature and finally the complete development of human predispositions. Kant's overall instrumentalist language in this context that depicts the human being as the object of nature's willing seems to even reassert this reading.[20] Although the ambiguity of Kant's exposition cannot be denied, this would be a misreading. For, as we will see further, with respect to culture "the human species" cannot be regarded just as "the foremost instrument for establishing order and consensus in irrational nature outside him" (CJ, 5:431). This is so because the human being is a *cultural* being in the specific sense that it has the capacity to set itself ends and to "us[e] nature as a means appropriate to

---

[19] In this paper the considerations on culture are restricted to the context of teleology and history. For a much broader discussion of culture and its differentiation into a narrow and wide meaning, cf. Klingner (2013, 19–60).

[20] Cf., for example, *Idea*, 8:21: "The human being wills concord; but nature knows better what is good for his species: it wills discord" (*Idea*, 8:21). See CJ, 5:430–431.

the maxims of his free ends in general" (CJ, 5:431). Thus, although the human being is subject to nature willing him or her to fully develop his or her natural dispositions and with the help of the drive to society to civilize and federalize, culture on the other hand elevates the human being above nature in that it uses it as a means for its own free purposes.

As a consequence, since according to the doctrine of the system of ends, the ultimate end of nature is indispensable and since it cannot be determined *contentwise*, it must be *formally* determined. The formal determination of the ultimate end of nature consists in the human being's "aptitude" to set ends and to seek in nature the appropriate means to achieve them. The ultimate end of nature "in regard to the human species" is therefore "culture" as "[t]he production of the aptitude of a rational being for any ends in general" (CJ, 5:431). "Aptitude" must be understood as the human being's capacity to appropriate nature for the realization of ends set by the human being herself. By using nature as means, the human being produces objects for her own purposes. In a broad sense, this is what "culture" and its formal condition, that is, the "aptitude" for this production, means.[21] The question is whether the coming into being of culture results as a byproduct of the realization of the social drive or whether there is in addition to the social drive a drive to *culture* that explains why in the course of human socialization culturalization takes place. It is highly plausible to assume that socialization and culturalization go hand in hand since social bonds urge the human being to produce what it needs in order to organize peaceful coexistence, be it industry, education or transport. But this is not analytically true, since it is not contradictory to imagine a society devoid of cultural achievements. This seems to suggest that the drive to society needs to be accompanied by what one might call the drive to culture, that is, another natural drive that motivates the human social being to produce the "aptitude" "for any ends in general" (CJ, 5:431). But Kant does not consider the drive to culture. He seems to believe that culturalization is a byproduct of the realization of the drive to society in nature. The existing gap between the natural propensity to socialize on the one side and the *free* production of that aptitude on the other remains largely unnoticed by him.

[21] Cf. Düsing (1968, 212f).

Kant rather integrates socialization and culturalization such that the drive to culture is an element of the social drive. The capacity to use nature as a means for the realization of ends is not simply present or given in the human being but must be formed under the social condition. This is accomplished in the history of the human species through the development of natural dispositions, which offers to the human being through freely setting ends and choosing the appropriate means the possibility of exceeding herself. The human being is not outside of culture but part of it under the universal conditions that nature presets. Here Kant differentiates between the culture of "skill" and the culture of "discipline" (CJ, 5:431–433). On the one hand, it is true that the human being needs the skill in order to reach her ends, but, on the other, skill alone is not sufficient to choose between possible ends. This choice requires, as Kant says, the culture of "discipline" (CJ, 5:432) by means of which the will is *disciplined* in the choice of ends. Like in the *Idea*, Kant explains the formation of skill with the help of the drive to society and the consequences of its realization: "Skill cannot very well be developed in the human race except by means of inequality among people" (CJ, 5:432). He thinks that the differences between individual capacities of the human beings and the conflicts among them further the development of natural dispositions in general and lead to cultural "progress": "yet this splendid misery is bound up with the development of the natural predispositions in the human race, and the end of nature itself, even if it is not our end, is hereby attained" (CJ, 5:432). The development of culture is therefore incomprehensible independently of the development of natural dispositions, especially of the drive to society. As the ultimate end of nature, culture is itself part of nature and is even produced by nature, at least indirectly thorough the drive to society. For nature realizes its ultimate end under one condition that constitutes the point on which the arguments of the *Idea* and the third *Critique* concur:

> The formal condition under which alone nature can attain this its final aim is that constitution in the relations of human beings with one another [...], which is called *civil society*; for only in this can the greatest development of the natural predispositions occur. For this, however, even if humans were clever enough to discover it and wise enough to subject themselves willingly

to its coercion, a *cosmopolitan* whole, i.e., a system of all states that are at risk of detrimentally affecting each other, is required. (CJ, 5:432)

Kant is convinced that natural dispositions can develop in the human species only under the formal political and juridical condition of a legal, freedom guaranteeing order, that is, civil society and ultimately the cosmopolitan federation of states. One might wonder why he believes this to be the case. Kant thinks that natural dispositions do not develop purposefully in the anarchic state of nature, since their formation is then hampered by the war of all against all. This does not mean that natural dispositions do not need for their evolution resistant forces in order to come to light and to further, through their development, the progress of culture. As the fourth proposition of the *Idea* explains, this is even an explicit requirement: "The means nature employs in order to bring about the development of all its predispositions is their antagonism in society, insofar as the latter is in the end the cause of their lawful order" (*Idea*, 8:20). While in the *Idea*, the question about the link between the development of natural predispositions and the drive to society remains open insofar Kant there simply presupposes that natural predispositions of human beings must develop completely and in conformity to a plan (*Idea*, 8:18 f.), the *Critique of the Power of Judgment* and its theory of the ultimate end of nature provides the subsequent teleological foundation for this hypothesis.

As we have seen, the theory of the ultimate end is based on the concept of external purposiveness and on the means-end relation that constitutes it. It is the external purposiveness of organized beings, in coordination with internal purposiveness, which leads to the ultimate end of nature. This ultimate end of nature is the human being as natural being motivated by means of the social drive not only to civilize but also to culturalize. In the framework of the system of ends for which the human being is the ultimate end of nature, Kant's claim is well-justified that the natural dispositions of the human being develop purposefully only in the species for as a rational being the human being itself not only chooses its ends but also the appropriate means to those ends. When setting ends, the human being as cultural being naturally aims at ensuring its existence and, consequently, at avoiding conflicts that threaten its safety. This end can be achieved no other than under the condition of the civil constitution and ultimately a federation of states. The only means to realize this end are natural dispositions, whose formation and development in history Kant conceives as culture. But independently of the drive to society culture would not occur.

5 Conclusion

At the outset, I have claimed that Kant proposes a conception of "drive" that integrates the natural and cultural dimension of human life. The drive to society is introduced in the philosophy of history as natural disposition whereas in the philosophy of biology in the narrow sense drive does not have a systematic place. As such Kant takes drive to be the means by which nature promotes socialization. Socialization is not a process that gradually evolves in the course of human history. An inherent feature of the social drive is the antagonism that nature deploys in order to further the complete development of dispositions. Independently of emerging social structures along with inherent conflicts, motivated through the drive to society, cultural achievements and ultimately the complete development of dispositions are not possible. On the face of it, Kant succeeds in combining the natural history of mankind with human social history in a way that integrates nature and culture. Like in transcendental philosophy proper, the claim Kant makes here is a modest one. For we may to a certain degree determine the history of humankind in terms of its biological coming into being apart from teleological considerations, but with respect to culture we can only conceive of the human being's history *as if* it is purposeful while not to being able to positively determine a plan that underlies human's historical development. For many Kant's teleology as the theory of intrinsic purposiveness of natural ends is a challenge and rather unattractive. In that respect Kant's philosophy of history and culture is even a bigger obstacle. For culture as the effect of the drive to society materializes only because internal purposiveness is combined with external purposiveness, that is, the civil and federalist condition of securing social coexistence. It can certainly not be denied that under peaceful conditions in a state or international federal bond, the flourishment of culture and, according to Kant, by implication the complete development of natural dispositions is more likely. Nevertheless, humans' *freely* setting ends, that is, culture, must still be considered as alien to these external conditions. Even if one accepts intrinsic purposiveness, it is still not easy to see how nature and culture ultimately converge with the help of the social drive. The drive to society is a *natural* predisposition that as such cannot explain how in the emergence of social structures the achievement of culture does overcome the Kantian dualism of nature and rationality. For culture is different from nature in that the human being makes nature a means for realizing ends that are not set by nature but that she freely sets herself. And freely setting ends is irreducible to the efficacy of nature and the *willing* that Kant attributes to it. As we have seen in Sect. 1, Kant inherits this limitation form the early modern conception of conceiving the drive to society in purely natural terms. In the aftermath the categorical separation of the natural from the rational requires its reconciliation, that is, in Kantian terms, the integration of natural predispositions and culture. Kant was aware that no positive cognitive account of how this might be theoretically possible can be given. Within the frame of reflective power of judgement, we can at most conceive of nature *as if* it is purposeful with respect to culture. Kant's separation of nature and rationality does therefore leave no room for second nature.

References

Allison, Henry E. 2009. Teleology and History in Kant: The Critical Foundations of Kant's Philosophy of History. In *Kant's Idea for a Universal History with a Cosmopolitan Aim: A Critical Guide*, ed. Amélie Oksenberg Rorty and James Schmidt, 24–45. Cambridge: Cambridge University Press.

Buchenau, Stefanie. 2002. Trieb, Antrieb, Triebfeder dans la philosophie morale prékantienne. *Revue germanique internationale* 18 (Trieb: tendance, instinct, pulsion): 1–24.

Crusius, Christian August. 1744. *Anweisung vernünftig zu leben*. Leipzig: Gleditsch.

Düsing, Klaus. 1968. *Die Teleologie in Kants Weltbegriff* (Kant-Studien Ergänzungsheft 96). Bonn: Bouvier.

Heidemann, Dietmar. 2010. Allgemeine Zweckmäßigkeit der Natur. In *Zweck und Natur: Historische und systematische Untersuchungen zur Teleologie*, ed. Tobias Schlicht, 91–111. München: Fink.

———. 2016. Letzte Zwecke. Die Grundlegung des Föderalismus in Kants politischer Teleologie. In *Join or Die: Philosophical Foundation of Federalism*, ed. D. Heidemann and K. Stoppenbrink, 93–111. Berlin/Boston: De Gruyter.

Kant, Immanuel. 1900 ff. *Gesammelte Werke*. Edited by the Royal Prussian (later German) Academy of Science. Berlin: De Gruyter.

———. 1998. *Religion within the Boundaries of Mere Reason*. Edited and translated by A. Wood, and G. di Giovanni. Cambridge: Cambridge University Press.

———. 1999. Groundwork of the Metaphysics of Morals. In *Immanuel Kant, Practical Philosophy*, edited and translated by M. J. Gregor. Cambridge: Cambridge University Press.

———. 2000. *Critique of the Power of Judgment*. Translated by P. Guyer, and E. Matthews. Cambridge: Cambridge University Press.

———. 2007a. Idea for a Universal History with a Cosmopolitan Aim. Translated by Allen W. Wood. In *Immanuel Kant, Anthropology, History, and Education*, ed. G. Zöller and R. B. Louden, 107–120. Cambridge: Cambridge University Press.

———. 2007b. Anthropology from a Pragmatic Point of View. Translated by Robert B. Louden. In *Immanuel Kant, Anthropology, History, and Education*, ed. G. Zöller and R. B. Louden, 227–429. Cambridge: Cambridge University Press.

Klingner, Stefan. 2013. *Technische Vernunft: Kants Zweckbegriff und das Problem einer Philosophie der technischen Kultur*. Berlin/Boston: De Gruyter.

Schneewind, Jerome B. 2009. Good Out of Evil: Kant and the Idea of Unsocial Sociability. In *Kant's Idea for a Universal History with a Cosmopolitan Aim: A Critical Guide*, ed. Amélie Oksenberg Rorty and James Schmidt, 94–111. Cambridge: Cambridge University Press.

Thomasius, Christian. 1692. *Von der Kunst Vernünfftig und Tugendhafft zu lieben: Als dem eintzigen Mittel zu einen glückselligen, galanten und vergnügten Leben zu gelangen: Oder: Einleitung Zur SittenLehre*. Halle (Saale): Salfeld.

———. 1696. *Von der Artzeney Wider die unvernünfftige Liebe und der zuvorher nöthigen Erkäntnüß Sein Selbst: Oder: Ausübung Der SittenLehre*. Halle (Saale): Salfeld.

9

Feeling and Life in Kant's Account of the Beautiful and the Sublime

Yoon H. Choi and Alix Cohen

1 Introduction

The aim of this chapter is to explore what Kant means by "life", the "feeling of life", the "feeling of the promotion (or inhibition) of life", and related notions, such as the idea of a "vital power".

Kant often points to acts of willing as manifestations of life. For example, in the *Critique of Practical Reason*, life is defined as "the faculty of a being to act in accordance with laws of the faculty of desire" (CPrR 144 [5:9n]; see also LM-L_2 346 [28:587], LM-Mrongovius 261 [29:894], and LM-Dohna 381 [28:680], inter alia). This leaves the impression that life and its attendant feelings are to be understood primarily, if not solely,

Y. H. Choi (✉)
Marquette University, Milwaukee, WI, USA
e-mail: yoon.choi@marquette.edu

A. Cohen
University of Edinburgh, Edinburgh, UK
e-mail: alix.cohen@ed.ac.uk

M. Kisner, J. Noller (eds.), *The Concept of Drive in Classical German Philosophy*,
https://doi.org/10.1007/978-3-030-84160-7_9

in relation to the faculty of desire. This, however, cannot be the whole story. For in an oft-quoted and rich description of the feeling of the beautiful and the feeling of the sublime, Kant says that

> the former (the beautiful) directly brings with it [*bei sich führt*] a feeling of the promotion of life [*Gefühl der Beförderung des Lebens*], and hence is compatible with charms and an imagination at play, while the latter (the feeling of the sublime) is a pleasure that arises only indirectly, being generated namely by the feeling of a momentary inhibition of the vital powers [*Lebenskräfte*] and the immediately following and all the more powerful outpouring of them. (CJ 128–129 [5:244–245])

To make sense of the connection between life and aesthetic experience, we set out an account of Kant's conception of life and the various feelings associated with it. We then use this analysis to guide our interpretation of the passage quoted above. We argue that it is significant that Kant characterizes the feeling of the beautiful as a feeling of the promotion of life but the feeling of the sublime in terms of vital powers. We account for this difference by showing that in experiences of beauty, we are aware of ourselves as rational but embodied human beings, as part of nature. In the feeling of the sublime, by contrast, we are aware of ourselves as pure rational beings. This entails that in the beautiful, we experience nature as deeply life-promoting, while in the sublime, we experience the independence of our supersensible vocation from nature.

2 Being Alive and Feeling Alive

In this section, we set out our interpretation of Kant's conception of life, discussing the relation between life and the "principle of life" before turning to the various feelings that pertain to life. In the *Critique of Pure Reason*, Kant says that when we represent the relation between ourselves as thinking beings and our bodies, we conceive of "the thinking substance as the principle of life in matter, i.e., as a soul (anima)" (CPR 414

[A345/B403]).[1] Though this passage appears in the Paralogisms, where the aim is to deny the rational psychologist's claim to knowledge of such thinking substances, Kant accepts the view that matter itself is lifeless, inert, and that it must be animated by a non-material, inner "principle of life" in order for life to obtain. This is true not only for the human being, in whom the "animating principle [...], the soul", not only thinks but is also "free, self-determining" (RT 434 [8:392–393]; see also TPP 456 [8:417]), but also for the non-human animal, which represents but does not think.[2]

Perhaps more surprisingly, Kant further suggests that life obtains *only* when the principle of life animates a body, when there is a "combination of soul with body":

> [I]t is impossible for us to know whether and what the life-principle in man (the soul) is able to do in thinking without the body, and everything here amounts only to empirical knowledge, i.e., to a knowledge that we can acquire in life, and hence in a combination of soul with body. (RP 376 [20:286])

Even if the soul-qua-principle of life is independent of the body, the capacity of the soul to think and act depends, as far as we know, on the body. Thus "were the soul separated from the body", as Kant reportedly put it in one of his metaphysics lectures, "its life power may indeed not stop, but its life would" (LM-Mrongovius 279 [29:914]).[3]

---

[1] One difficulty with working out Kant's views on this and related points is that, although there is a large amount of relevant textual evidence, most of it comes in the form of student lecture notes. Moreover, these notes indicate that Kant's views on these matters were continuously evolving. We have tried to mitigate these difficulties when possible by quoting from texts that were either published or written to be published and by citing several supporting sources. For a helpful survey of the shifts in Kant's views, see Caygill (2000). For a detailed examination of the historical roots of Kant's concept of life, see Fugate (2018).

[2] We thus take something to be alive if it "has a faculty to alter its own state as a consequence of its own representations" (LM-Volckmann 295 [28:448–449]). It remains a matter of debate, however, whether we are licensed to say that organisms really are alive. We bracket this issue, focusing only on life and the feeling of life as it obtains in human beings. For discussion, see McLaughlin (1990), Ginsborg (2001), Breitenbach (2013b, 2014).

[3] See also CPR 438 [A393], LM-Volckmann 354 [28:594]; LM-L_2 354 [28:594] and LM-K_2 396 [28:755] and 405 [28:675]. Cf. MFNS 251–252 [4:544].

If we put these claims together, the view that emerges is that life manifests in those activities that originate in and express our nature as embodied but rational beings, that is, as human beings. Life thus manifests not in the pure law-giving of pure reason, but in our feeling of respect; not in our brute affection by outer objects, but in our representing the object in intuition. More generally, since the "three major powers of the [human] soul"—the cognitive faculty, the faculty of desire (the will), and the feeling of pleasure and displeasure—"constitute its life" (LM-Mrongovius 247 [29:878]), life manifests not only in our cognitive activities and our exercises of moral and practical agency but also in the feelings that accompany our activities.[4]

Understanding life to obtain when and insofar as a principle of life animates a body allows us to see how there could be a *feeling* with respect to life at all. For if the fact that the *principle of life* in us human beings is a supersensible principle of freedom meant that human *life* had to be understood just as a supersensible capacity for free self-determination, how could there be a *feeling* of life at all? Isn't the domain of the supersensible inaccessible to sensibility? By contrast, if life is understood to obtain when a principle of life animates a body, there is no difficulty accommodating the possibility of feeling.

Understanding life as an essentially embodied phenomenon also helps make sense of the way Kant talks about the feeling of life. For Kant sometimes writes as if there is no such thing as *the way that life itself feels*. A set of anthropology lecture notes from the early Critical years, for instance, records Kant as saying, "[l]ife in itself cannot be felt, but only its promotion or hindrance" (LA-Mrongovius 427 [25:1319]). A similar view is suggested in the following passage from the *Third Critique*:

> [L]ife without the feeling of the corporeal organ is merely consciousness of one's existence, but not a feeling of well- or ill-being, i.e., the promotion or inhibition of the powers of life; because the mind for itself is entirely life (the principle of life itself), and hindrances or promotions must be sought

[4] As we mentioned earlier, Kant sometimes appears to identify life with the faculty of desire. (See Deimling (2018) for a careful defense of these claims.) We suggest, however, that desire is a sufficient condition or "mark" of life.

> outside it, though in the human being himself, hence in combination with his body. (CJ 159 [5:277–278])

If "life in itself" and "life without the feeling of the corporeal organ" can be understood as referring to the supersensible principle of life by itself, we can make sense of why Kant would deny that there is an associated feeling. In other passages, however, Kant suggests that there is indeed a feeling of life. These passages, significantly, speak of the feeling of life in the singular. The very first section of the *Analytic of the Beautiful*, for instance, says that in a judgment of beauty,

> the representation [of the beautiful object] is related entirely to the subject, indeed to its feeling of life, under the name of the feeling of pleasure or displeasure, which grounds an entirely special faculty for discriminating and judging that contributes nothing to cognition but only holds the given representation in the subject up to the entire faculty of representation, of which the mind becomes conscious in the feeling of its state. (CJ 90 [5:204]; see also CJ 158–159 [5:277])[5]

Though there is much in this passage that needs unpacking, we draw attention only to what is implied by connecting the "feeling of life" with the "entire faculty of representation". This suggests that the "feeling of life" is the overall feeling the human being has of herself as a unified and embodied rational agent. In accordance with the interpretation we adopt of Kantian feelings, we take the feeling of life to be an affective appraisal of the condition of the human being as a unified and embodied rational agent.[6] It is thus a specific and special feeling of pleasure and displeasure.

This brings us to the feeling of the promotion or inhibition of life. Given that life manifests in activities that originate in and express our nature as rational agents, something that facilitates these activities is felt

---

[5] See also the references to the "entire power of mind" (LM-L_1 63 [28:247]) and our "entire faculty" (LM-Mrongovius 258–259 [29:890–891]), as well as the reference to life as unity in an interesting *Reflexion* (R6862) that Adickes dates between 1776 and 1789 (NF 443 [19:183]).

[6] We adopt and build on an interpretation of the function of feeling in the Kantian mind elaborated and defended in earlier work by Cohen (2020, 2021).

as promoting of life and a pleasure. As Kant puts it, "[w]hat promotes our life, i.e., what brings our activity into play, as it were, pleases" (LL-Jäsche 31 [9:45]). By contrast, something that hinders these activities is felt as a hindrance to life and a displeasure (PA 334 [7:231], LL-Blomberg 31 [24:45], LA-Friedländer 117 [25:559], LM-Mrongovius 261 [29:894] and LM-L_2 346 [28:586], inter alia).

The last term we introduce is that of "vital powers", which Kant appeals to in characterizing the sublime as the "feeling of a momentary inhibition of the vital powers [*Lebenskräfte*] and the immediately following and all the more powerful outpouring of them" (CJ 128–129 [5:244–245]). Is the feeling of the "vital power" to be identified with the "feeling of life"? Or is it something different? Though Kant's usage is not entirely regimented, the term "vital powers" is generally used to refer to capacities considered independently of our rational nature. Kant connects it to his theory of temperament: "temperament", he writes, is something that "nature makes of the human being", something "where the subject is for the most part passive" (PA 390 [7:292]). Our temperament is thus determined by the characteristic way we use our "vital force": our power "insofar as it does not belong to the rationality of the human being" (LA-Busolt 523 [25:1531]; see also PA 385 [7:287]).[7]

Though this remains but a sketch of the concept of life and the various feelings related to it, a picture begins to emerge according to which, although we are alive in virtue of the supersensible principle of life in us, we live our lives as embodied beings and we feel alive in virtue of being embodied beings whose activities are alternatively facilitated and frustrated by the things we encounter in nature, including our sensible selves. We turn now to the feeling of the beautiful and the feeling of the sublime. The first, Kant says, brings with it the feeling of the promotion of life; the second, by contrast, is characterized in terms of our vital powers. In the remainder of this chapter, we aim to make sense of these claims.

[7] For example, Kant regularly uses "vital power" in discussing temperament but *not* character. See PA 384 ff. [7:285 ff.], LA-Mrongovius 467 [25:1370], and LA-Busolt 523–524 [25:1531], for example. See Cohen (2017b) for discussion.

3 The Experience of Beauty

According to Kant, the experience of beauty involves a feeling of aesthetic pleasure and the harmonious free play of imagination and understanding. Kant's account of beauty is famously controversial, and numerous interpretations of it have been put forward in the literature.[8] As the aim of this chapter is to attain a better understanding of Kant's conception of life and the feeling of life, this section focuses on making sense of Kant's claim that the beautiful "directly brings with it [*bei sich führt*] a feeling of the promotion of life [*Gefühl der Beförderung des Lebens*]" (CJ 128 [5:244]). We open with a brief comparison of aesthetic and cognitive judgment, characterizing the different kinds of mental activity involved. This helps us underscore the distinctiveness of aesthetic judgment.[9] We go on to take a closer look at the activity of harmonious free-play characteristic of experiences of beauty and the feelings that accompany it. We conclude that the feeling of pleasure elicited in our encounters with beautiful objects is the feeling of the promotion of life par excellence, as in these experiences we find the cognitive capacities central to our rational but embodied existence enhanced.

According to Kant, the faculties that are centrally involved in cognition—imagination and understanding—are also the faculties that enable the experience of beauty, but the imagination functions in very different ways in the two cases. In cognitive judgment, it composes the manifold of intuition under the direction of the understanding: its activity is unfree, but therefore lawful and harmonious with the activity of the understanding (see CJ 124–125 [5:240–241]). In aesthetic judgment, by contrast, the imagination, though it engages with formal features of the object of intuition and so is not engaged in pure "invention", nevertheless composes the manifold at will, freely: it is the "authoress of voluntary [*willkührlicher*] forms of possible intuitions" (CJ 124–125 [5:240–241]). Since "the understanding alone gives the law", imagination cannot, in its

[8] For example, see Guyer (1997, 2006, 2009), Allison (2001), Zuckert (2007), and Ginsborg (2015).

[9] By "cognitive judgments", we mean the range of judgments ultimately aiming at determinate concept-application: determining judgments but also reflecting judgments that aim at generating a systematic unity of empirical concepts and laws.

free activity, be "lawful by itself". When its activity nevertheless harmonizes with that of the understanding, it manifests a "lawfulness without law" (CJ 125 [5:241]). On our view, the difference in the activity of the imagination in the two cases leads to a difference in the kind of harmonious relation that obtains between the imagination and the understanding. In aesthetic experience, the harmonious relation that obtains is one in which "the understanding is in the service of the imagination" (CJ 126 [5:242]). It is thus a "subjective unity" that arises between imagination and understanding, a unity we can only be conscious of "aesthetically, through mere inner sense and sensation" (CJ 103 [5:218]).

On the view we just laid out, the activity of imagination and understanding in aesthetic experience is understood to be thoroughly different from the activity of these faculties in cognitive experience, and this difference stems from the freedom of imagination's activity in aesthetic experience.[10] As Kant puts it, "[e]verything flows from the concept of taste as a faculty for judging an object in relation to the free lawfulness of the imagination" (CJ 124 [5:240]; see also CJ 153 [5:270]). In what follows, we go on to identify two further features of aesthetic experience that we take to be made possible by the distinctive activity of imagination in experiences of beauty.

Kant says that a beautiful object draws the imagination into an activity that, "*in its freedom*[,] arouse[s] the understanding", while at the same time eliciting from the understanding an activity that, "without concepts, sets the imagination into a regular play" (CJ 175 [5:295–296]; emphasis added). According to this passage, the activity of the understanding and the activity of the imagination sustain each other in an interplay that intrinsically "strengthens and reproduces" itself; we thus find ourselves "linger[ing] over the consideration [*Betrachtung*] of the beautiful" (CJ 107 [5:222]). That is, the freedom of the imagination's activity plays a key role in accounting for the self-sustaining nature of the free play. Kant

[10] Though we cannot give a full defense of this reading of aesthetic and cognitive judgments here, the claims to follow do not depend on adopting this reading. For an alternative interpretation that identifies a "common core" between aesthetic and cognitive judgments, see Breitenbach (2013a, 2015, 2020, Forthcoming). The view we adopt has, on the face of it, a harder time accounting for the universality of aesthetic judgments (though see Cohen 2017a, 146–147), whereas a view such as Breitenbach's must deal with the prima facie counterintuitive implication that all cognizable objects are (or could be) beautiful.

further suggests that this self-sustaining feature makes aesthetic experiences cognitively enhancing. As he puts it, a "spirited" work of art "purposively sets the mental powers into motion, i.e., into a play that is self-maintaining and even strengthens the powers to that end"—the very end, that is, of being self-maintaining (CJ 192 [5:313]).[11] This suggests that Kant thinks a work of art may, by eliciting an activity that is inherently self-sustaining, induce the imagination and understanding to activity of uncommon intensity and duration. This, in turn, increases our cognitive stamina and gives us a glimpse of the degree of sustained and focused activity our cognitive capacities are capable of. It also allows us to experience the fact that we can find it deeply pleasurable to put our cognitive capacities to intensive use. Thus experiences of beauty leave us keener and abler users of our cognitive capacities.

Experiences specifically of natural beauty are also cognitively enhancing in a further way. Kant writes that "nature has the property of containing an occasion for us to perceive the inner purposiveness in the relationship of our mental powers in the judging of certain of its products", namely, the beautiful ones (CJ 224 [5:350]; see also CJ 103 [5:218]). The reference to inner purposiveness points to the harmonious relation that obtains between imagination and understanding when they are engaged in free activity. But what is this "relationship of our mental powers" purposive *for*? Kant writes:

> A representation which, though singular and without comparison to others, nevertheless is in agreement with the condition of universality, an agreement that constitutes the business of the understanding in general, brings the faculties of cognition into the well-proportioned disposition

[11] Although in this passage, Kant is talking about works of art, rather than natural beauty, the specific kind of cognitive enhancement characterized here seems to be secured just by the nature of the harmonious free play and thus seems generalizable to all beautiful experiences. The same is not true with respect to the kind of cognitive enhancement Kant attributes to the experience of beautiful poetry, which, according to Kant, "expands the mind by setting the imagination free and presenting, within the limits of a given concept and among the unbounded manifold of forms possibly agreeing with it, the one that connects its presentation with a fullness of thought to which no linguistic expression is fully adequate" (CJ 203–204 [5:326]). Though the freedom of imagination's activity remains central to this cognitive enhancement, the involvement of specific concepts raises the possibility that the sort of cognitive enhancement described here might be unique to the case of poetry.

> that we require for all cognition and hence also regard as valid for everyone (for every human being) who is determined to judge by means of understanding and sense in combination. (CJ 104 [5:219])

What is "the business of the understanding in general", and what is "the well-proportioned disposition that we require for all cognition"? In the *Prolegomena to Any Future Metaphysics*, the idea of a "consciousness in general" appears in several key passages and is clearly meant to be understood in contrast to the "consciousness of my state". By a "consciousness in general", Kant means a consciousness that attains to universal validity, rather than having merely subjective validity (P 94 [4:300]; see also P 98 [4:304] and 105 [4:312] and CPR 252 [B143]). If we understand "the business of the understanding in general" accordingly, we are led to view experiences of the beautiful as bringing our faculties into a relation that is purposive for attaining universal validity. They accordingly foster a "disposition" that is conducive to cognitive activity: a disposition to look for harmony to arise between what is essentially particular—beautiful objects—and what is universal (understanding in general), rather than clinging to either the particular or the universal and forcing an agreement between the two (CJ 104 [5:219]).

We have argued that the harmonious free play of imagination and understanding is an activity that is cognitively enhancing. Recall that we also argued in Sect. 1 that the feeling of the promotion of life is a feeling of pleasure elicited when we encounter something that promotes the activities that originate in and are expressive of our nature as embodied but rational agents.[12] It is thus because our cognitive capacities are capacities that are central to our rational and embodied nature that the feeling of pleasure elicited in our encounters with beautiful objects is a feeling of the promotion of life. Indeed, it is the feeling of the promotion of life par excellence. For the capacities involved in experiences of beauty are the capacities that make us the specific kind of rational being we are, namely, human beings, living in and sharing a spatiotemporally ordered world. Kant makes this clear when he writes that "beauty is valid only for human

[12] Cf. Zammito (1992, 295), whose analysis of the feeling of life focuses on ourselves in the "world of sense," and Makkreel (1990, 92), whose account connects the feeling of life with "pure mental spontaneity". We take our view to articulate a third option.

beings, i.e., animal but also rational beings, but not merely as the latter (e.g., spirits), rather as beings who are at the same time animal" (CJ 95 [5:210]). Feelings of beauty accordingly engage and promote those capacities that make us the kinds of beings we are and sustain our existence as the kinds of beings we are.

4 The Experience of the Sublime

In this section, we turn to the feeling of the sublime. We argue that the experience of the sublime, like the experience of beauty, occasions a special awareness of our rational capacities, and this, in turn, leads to their enhancement. But whereas the experience of beauty engages the imagination and understanding, the experience of the sublime draws in the imagination and reason. And whereas the feeling of aesthetic pleasure "directly brings with it" a feeling of the promotion of life, the sublime elicits "the feeling of a momentary inhibition of the vital powers and the immediately following and all the more powerful outpouring of them" (CJ 128–129 [5:244–245]). In this section, we analyze the experience of the sublime, arguing that it enables the awareness and thereby the enhancement of our rational capacities.[13] This sets us up to explain the significance of Kant's characterization of the feeling of the sublime in terms of "vital powers" rather than a "feeling of the promotion of life".

Whereas the feeling of the beautiful is "immediately pleasant", the feeling of the sublime is not; it begins with an unsettling and painful feeling of inadequacy. The experience of the mathematical sublime, for example, begins when the imagination, presented with a large sensible object, struggles to grasp it as a whole, to comprehend it aesthetically, "in one intuition" (CJ 138 [5:254]). In the case of the dynamical sublime, a display of the overwhelming power of nature makes plain to us that, as natural beings, our "capacity to resist" is an "insignificant trifle" (CJ 144 [5:261]). It is important to note, however, that this initial painful moment

[13] See Matthews (1996) and Clewis (2009) for helpful analyses of the sublime; both, however, focus on the role of the sublime in enabling awareness of our supersensible capacities, whereas we argue that the experience of the sublime also enhances our ability to use these capacities.

in the encounter with the sublime is conditioned by our "receptivity to ideas" of reason, which can make "what is repellent for the sensibility [...] at the same time attractive for it [the sensibility]" (CJ 148 [5:265]; see also 172 [5:292]). Thus in the case of the mathematical sublime, imagination seeks to grasp the large object as a whole because it is answering a demand for totality and wholeness that originates in reason. As Kant puts it, "[T]he idea of the comprehension of every appearance that may be given to us into the intuition of a whole is one enjoined on us by a law of reason" (CJ 140 [5:257]). And in the case of the dynamical sublime, it is our receptivity to the moral ideas of reason that makes threatening vistas compelling and draws us toward them. In the absence of such ideas, Kant says, such vistas would be "merely repellent" (CJ 148 [5:265]; see also CJ 150 [5:267], 151 [5:268], and 129 [5:245–246]).

Though reason thus plays an enabling role in the experience of the sublime, the initial moment of the experience itself, as we detailed earlier, is constituted by the painful awareness of our limitations as imagination is driven "as it were to an abyss, in which it fears to lose itself" (CJ 141 [5:258]). But "displeasure", according to Kant, "is that representation that contains the ground for determining the state of the representations to their own opposite (hindering or getting rid of them)" (CJ 105 [5:220]). The painful awareness of our inadequacy thus involves a strong impetus to retreat and disengage; this is what we take to manifest as the "feeling of a momentary inhibition of the vital powers" (CJ 128–129 [5:244–245]). Yet when we are caught up in an experience of the sublime, we need only attempt to withdraw and we find our retreat "immediately" checked by the superior attractive pull of what drew the imagination toward the abyss in the first place: reason and our supersensible vocation. Accordingly, the "feeling of a momentary inhibition of the vital powers" gives way to "the immediately following and all the more powerful outpouring of them" as we are held fixed, our flight to safety arrested by our own attraction to the supersensible, our sensible faculty of imagination left "expend[ing]" itself at the behest of reason (CJ 128–129 [5:244–245]). To experience this, according to Kant, is to feel what it is like "to abandon sensibility" for the sake of "ideas that contain a higher purposiveness" (CJ 129 [5:245–246]); it is to experience and find "palpable [...] the sublimity of its [the mind's] own vocation even over nature" (CJ 145 [5:262]).

On our analysis of the experience of the sublime, the effort of the imagination and the feeling it gives rise to is contrapurposive and painful for our sensible nature, but purposive and pleasurable in relation to the "vocation of the mind that entirely oversteps the domain of the [sensible]", namely our supersensible vocation (CJ 151 [5:269]). As Kant puts it,

> [T]he inner perception of the inadequacy of any sensible standard for the estimation of magnitude by reason corresponds with reason's laws, and is a displeasure that arouses the feeling of our supersensible vocation in us, in accordance with which it is purposive and thus a pleasure to find every standard of sensibility inadequate for the ideas of the understanding. (CJ 141 [5:258])[14]

Kant hints at a number of ways in which the experience of the sublime has an enhancing effect on our use of our rational capacities. The imagination, for example, is described as the "greatest faculty of sensibility" (CJ 141 [5:257]), for in its activity of apprehension, it "advances to infinity" (CJ 137 [5:253]; see also CJ 135 [5:251–252]).[15] But in the experience of the mathematical sublime, imagination's free acquiescence with the demands of reason effects an "enlargement of the imagination" (CJ 133 [5:249]; see also CJ 139 [5:256], 141–142 [5:258], and 148 [5:265]). As "an instrument of reason" (CJ 151 [5:269]), imagination feels itself "unbounded precisely because of [the] elimination of the limits of sensibility" (CJ 156 [5:274]; see also CJ 138 [5:255]). And this, in turn, "expands the soul" (CJ 156 [5:274]). It is not entirely clear how we are to

[14] Though the passage quoted earlier focuses on the mathematical sublime, the same point is made explicitly with reference to both the mathematical and the dynamical sublime in the following passage: "Reason inevitably comes in as a faculty of the independence of the absolute totality, and produces the efforts of the mind, though it is in vain, to make the representation of the senses adequate to that. This effort, and the feeling of the unattainability of the idea by means of the imagination, is itself a presentation of the subjective purposiveness of our mind in the use of the imagination for its supersensible vocation" (CJ 151 [5:268]).

[15] It is unable, however, to grasp and comprehend everything it can apprehend (CJ 135 [5:251–252] and 138 [5:255–256]).

understand these claims, but there are at least two different possibilities hinted at by the text. The first characterizes soul-expanding experiences as sweeping away "limits [...] arbitrarily set for [us]" that encourage us to the passive use of reason (CJ 156 [5:274–275]). By removing inducements to passive thinking, the enlargement of imagination may enhance our ability to think for ourselves (see CJ 174 [5:294]). A second possibility is suggested by Kant's observation that an imagination that proceeds "in accordance with the law of association" presents our "state of contentment [as] physically dependent", whereas as an "instrument of reason", imagination can represent things in a manner that "assert[s] our independence in the face of the influences of nature" (CJ 152 [5:269]). This suggests that experiences of the sublime leave behind an imagination that can better—more creatively, perhaps more courageously—represent the moral possibilities that arise in the world.

Experiences of the sublime also enhance our capacities by providing us with a corrected and sharpened awareness of the strength of our supersensible capacities. Indeed, these experiences make our supersensible powers downright "palpable" (CJ 145 [5:262]), though only through what Kant calls "negative presentation[s]" (CJ 156 [5:274]). Recall that the experience of the sublime begins with the humbling of "imagination in all its boundlessness", an experience we find deeply painful (CJ 140 [5:257]; see also CJ 145 [5:262]). As we understand Kant, the painfulness of the first moment affectively registers the overwhelming greatness and power of nature. But it also serves as a sensible measure of the superiority of reason over sensible nature as the latter is reduced to negligible significance in relation to the former. It is thus through the experience of the inadequacy of our sensible nature that the superiority of reason becomes "as it were [...] intuitable" to us (CJ 141 [5:257]). Similarly, displays of the ferocity of nature's power may expose our utter physical powerlessness, but in doing so, they expose the powerlessness of even such a dominating force to threaten "the humanity in our person" (CJ 145 [5:262]). In this manner, our experience of "the irresistibility of [nature's] power" ultimately serves to "make palpable" the power in us that nature cannot in the least threaten: the power that gives us our "highest principles" and our supersensible vocation (CJ 145 [5:261–262]).

Thus both the experience of the mathematical sublime and the experience of the dynamical sublime give us a better sense of the nature and strength of our rational capacity. But there is an important difference between the two cases. In experiencing the mathematical sublime, "the subject's own incapacity", Kant says, "reveals the consciousness of an *unlimited* capacity of the very same subject" (CJ 142 [5:259]; emphasis added). Importantly, he is referring specifically to reason's actualized capacity to think the absolute, which drives our cognitive activities and is necessarily unlimited (see CJ 142 [5:259] and 143 [5:260]). By contrast, what we become aware of as absolutely superior to nature in our experience of the dynamical sublime is that in us which issues our moral vocation, that in us which issues the *demand* that we develop and actualize our moral capacities. As Kant notes, "[T]he satisfaction here concerns only the vocation of our capacity as it is revealed to us [...] while the development and exercise of it is left to us and remains our responsibility" (CJ 145–146 [5:262]; see also CJ 154 [5:271]). The experience of the sublime nevertheless helps us with this task, too, for the heightened awareness of the strength of our supersensible capacities and their independence from nature "gives us the courage [*Mut*] to measure ourselves against the apparent all-powerfulness of nature" (CJ 144 [5:261]; see also CJ 151 [5:268]). The better our awareness of the strength of our supersensible powers, the better our awareness of the weakness of sensible nature, including our sensible inclinations, which can begin to appear more appropriately negligible to us than they normally do. This does not make us any "freer" from the determination of desires than we already are; nor does it make us more rational or virtuous than we already are. Rather, it enhances the way we exercise our powers as well as our determination to use them to the best of our ability. As Kant puts it, encounters with the dynamical sublime "leave behind a disposition of mind that, even if only indirectly, has influence on the consciousness of its strength and resolution in regard to that which brings with it intellectual purposiveness" (i.e., the supersensible) (CJ 155 [5:273]).[16]

[16] Though this passage is about the dynamical sublime, Kant also characterizes experiences of the mathematical sublime as producing a "disposition of the mind" useful for our practical pursuits; see CJ 139 [5:256].

But if, as we have argued, the experience of the sublime is one that enhances our ability to use our rational capacities, why is it characterized in terms of feelings of our vital powers rather than as a feeling of the promotion of life? Recall that Kant often speaks of "vital powers" when he is conceiving of our capacities as wrought by nature. In an experience of the sublime, we are aware of ourselves as moral beings whose lives and identities as embodied rational agents are secondary to our moral identities. Moreover, we experience what it is to be able to renounce our natural lives and identities to preserve our moral identities. As Kant puts it, the experience of the sublime "calls forth our power (which is not part of nature) to regard those things about which we are concerned (goods, health *and life*) as trivial" (CJ 145 [5:262]; emphasis added). We thus realize we are able to view the concerns, activities, and projects that constitute our embodied lives as akin to vital powers: ultimately inessential to our continued moral existence and thus expendable without loss of identity.

5 Conclusion

We have argued that both the experience of the sublime and the experience of the beautiful enhance our rational capacities. But while the experience of the beautiful counts as a feeling of the promotion of life, the experience of the sublime is characterized in terms of our vital powers. To conclude, we explore this difference a bit further.

According to Kant:

> [f]or the beautiful in nature we must seek a ground outside ourselves, but for the sublime merely one in ourselves and in the way of thinking that introduces sublimity into the representation of [nature]. (CJ 130 [5:246])

But in both the experience of the beautiful and the experience of the sublime, there is a sense in which the natural object or vista that we perceive serves only to catalyze the aesthetic experience. So why is the "beautiful in nature" grounded "outside ourselves"? Why is it that, as Kant writes, "we express ourselves on the whole incorrectly if we call some object of nature sublime, although we can quite correctly call very many of them beautiful?" (CJ 129 [5:245]). One part of the answer to this question lies in the fact that "what is properly sublime [...] concerns only ideas of reason" (CJ 129 [5:245]), and such ideas go "beyond the possibility of experience" (CPR 399 [A320/B377]). Another part of the answer lies in the fact that when we find a natural object beautiful, we find that it is "mere reflection" on its form that brings pleasure; so "[t]he ground of the pleasure is placed merely in the form of the object" (CJ 76 [5:190]). The vista or natural object that elicits the sublime, by contrast, "can be considered as entirely formless or shapeless" (CJ 160 [5:279]), since it is a "limitlessness" in what is perceived that elicits the experience of the sublime (CJ 128 [5:244]).[17] That we represent the beautiful as grounded "outside ourselves" but the sublime as grounded "in ourselves" (CJ 130 [5:246]) points to the fact that in our experience of the sublime, we are aware of ourselves as independent of nature. Such awareness cannot be grounded in nature but only in ourselves. In our experience of the beautiful, by contrast, we are aware of ourselves as living in nature, as part of nature. Our faculty of understanding, though legislative for experience, cannot function independently of sensibility; its dependence on sensibility means that it can be enhanced or inhibited in its pursuits by what is given in sensible nature. We thus find the experience of harmony obtaining between our sensible faculty of imagination and our supersensible faculty of understanding fortuitous.

What our view suggests is that aesthetic experiences are not only unique to human beings but also most representative of us. In the experience of the beautiful, we are aware of ourselves as part of nature, as living in nature, and, gratifyingly, as at home in nature. In our experience of the sublime, by contrast, we are aware of our supersensible nature. We feel our own idiosyncratic sensible lives recede and become mere nature in relation to our moral identities. It is significant that we cannot unify these two aesthetic experiences, but it is also significant that we can find them both pleasurable and enhancing.

[17] A complete answer must further reckon with the murky claims about the "supersensible substratum" that unifies theoretical reason and practical reason and enables the transition from nature to freedom. See CJ 227 [5:353] and 216–217 [5:340–341].

Abbreviations used: References to Kant's works use the abbreviations below, followed first by the page number from the translation consulted (see bibliography) and then the volume and page number from the *Akademie* edition in square brackets (Kant, Immanuel. 1902 ff. *Kants Gesammelte Schriften*, ed. Prussian Academy of Sciences. Berlin: De Gruyter).

CPR: *Critique of Pure Reason* (1781/1787), Kant (1998)
CPrR: *Critique of Practical Reason* (1788), Kant (1999)
CJ: *Critique of the Power of Judgment* (1790), Kant (2000)
LA: *Lectures on Anthropology*, Kant (2014)
LL: *Lectures on Logic*, Kant (1992)
LM: *Lectures on Metaphysics*, Kant (1997)
MFNS: *Metaphysical Foundations of Natural Science* (1786), Kant (2002)
NF: *Notes and Fragments*, Kant (2005)
P: *Prolegomena to Any Future Metaphysics* (1783), Kant (2002)
PA: *Pragmatic Anthropology* (1798), Kant (2007)
RP: "What Real Progress Has Metaphysics Made in Germany Since the Time of Leibniz and Wolff?" (1793/1804), Kant (2002)
RT: "On a Recently Prominent Tone of Superiority in Philosophy" (1796), Kant (2002)
TPP: "Proclamation of the Imminent Conclusion of a Treaty of Perpetual Peace in Philosophy" (1796), Kant (2002)

Acknowledgments The authors would like to thank the editors, Manja Kisner and Jörg Noller, for their patience and ongoing support, as well as all the participants of the Munich conference for their helpful feedback, with special thanks to John Zammito for his insightful comments. We are also grateful to Angela Breitenbach, Robert Clewis, and Ido Geiger for the time they so generously spent discussing some of the ideas developed in the paper.

References

Allison, Henry E. 2001. *Kant's Theory of Taste: A Reading of the* Critique of Aesthetic Judgment. Cambridge: Cambridge University Press.
Breitenbach, Angela. 2013a. Aesthetics in Science: A Kantian Proposal. *Proceedings of the Aristotelian Society* 113: 83–100.

———. 2013b. Kant on Biology and the Experience of Life. In *Kant und die Philosophie in weltbürgerlicher Absicht: Akten des XI. Kant-Kongresses 2010*, ed. Margit Ruffing, Claudio La Rocca, Alfredo Ferrarin and Stefano Bacin, 19–30. Berlin: De Gruyter.

———. 2014. Biological Purposiveness and Analogical Reflection. In *Kant's Theory of Biology*, ed. Eric Watkins and Ina Goy, 131–148. Berlin: De Gruyter.

———. 2015. Beauty in Proofs: Kant on Aesthetics in Mathematics. *European Journal of Philosophy* 23 (4): 955–977.

———. 2020. One Imagination in Experiences of Beauty and Achievements of Understanding. *British Journal of Aesthetics* 60: 71–88.

———. Forthcoming. Imaginative Reflection in Aesthetic Judgment and Cognition. In *Proceedings of the 13th International Kant Congress*, ed. Camilla Serck-Hanssen and Beatrix Himmelmann. Berlin: De Gruyter.

Caygill, Howard. 2000. Life and Aesthetic Pleasure. In *The Matter of Critique: Readings in Kant's Philosophy*, ed. Andrea Rehberg and Rachel Jones, 79–92. Manchester: Clinamen Press.

Clewis, Robert. 2009. *The Kantian Sublime and the Revelation of Freedom.* Cambridge: Cambridge University Press.

Cohen, Alix. 2017a. Kant on Beauty and Cognition: The Aesthetic Dimension of Cognition. In *Thinking about Science and Reflecting on Art: Bringing Aesthetics and the Philosophy of Science Together*, ed. Steven French and Otávio Bueno, 140–154. London: Routledge.

———. 2017b. The Natural, the Pragmatic and the Moral in Kant's Anthropology: The Case of Temperaments. *Early Science and Medicine* 22 (2–3): 253–270.

———. 2020. A Kantian Account of Emotions as Feelings. *Mind* 129 (514): 429–460.

———. 2021. Feeling, Orientation and Agency in Kant: A Response to Merritt and Eran. *Kantian Review* 26 (3): 379-391.

Deimling, Wiebke. 2018. Two Different Kinds of Value? In *Kant and the Faculty of Feeling*, ed. Kelly Sorensen and Diane Williamson, 25–41. Cambridge: Cambridge University Press.

Fugate, Courtney. 2018. Life and Kant's *Critique of Aesthetic Judgment.* In *Recht und Frieden in der Philosophie Kants: Akten des X. Internationalen Kant-Kongresses,* ed. Valerio Rohden, Ricardo R. Terra, and Guido A. de Almeida, 609–621. Berlin: De Gruyter.

Ginsborg, Hannah. 2001. Kant on Understanding Organisms as Natural Purposes. In *Kant and the Sciences*, ed. Eric Watkins, 231–258. Oxford: Oxford University Press.

———. 2015. *The Normativity of Nature: Essays on Kant's* Critique of Judgment. Oxford: Oxford University Press.

Guyer, Paul. 1997. *Kant and the Claims of Taste*. Cambridge: Cambridge University Press.

———. 2006. The Harmony of the Faculties Revisited. In *Aesthetics and Cognition in Kant's Critical Philosophy*, ed. Rebecca Kukla, 162–193. Cambridge: Cambridge University Press.

———. 2009. The Harmony of the Faculties in Recent Books on the *Critique of the Power of Judgment*. *Journal of Aesthetics and Art Criticism* 67 (2): 201–221.

Kant, Immanuel. 1992. *Lectures on Logic*. Edited and translated by J. Michael Young. Cambridge: Cambridge University Press.

———. 1997. *Lectures on Metaphysics*. Edited and translated by Karl Ameriks and Steve Naragon. Cambridge: Cambridge University Press.

———. 1998. *Critique of Pure Reason*. Edited and translated by Paul Guyer and Allen Wood. Cambridge: Cambridge University Press.

———. 1999. *Practical Philosophy*. Edited and translated by Mary J. Gregor. Cambridge: Cambridge University Press.

———. 2000. *Critique of the Power of Judgment*. Edited by Paul Guyer and translated by Paul Guyer and Eric Matthews. Cambridge: Cambridge University Press.

———. 2002. *Theoretical Philosophy after 1781*. Edited by Henry Allison and Peter Heath and translated by Gary Hatfield, Michael Friedman, Henry Allison, and Peter Heath. Cambridge: Cambridge University Press.

———. 2005. *Notes and Fragments*. Edited by Paul Guyer and translated by Curtis Bowman, Paul Guyer, and Frederick Rauscher. Cambridge: Cambridge University Press.

———. 2007. *Anthropology, History and Education*. Edited by Günter Zöller and Robert Louden and translated by Mary Gregor, Paul Guyer, Robert Louden, Holly Wilson, Allen Wood, Günter Zöller, and Arnulf Zweig. Cambridge: Cambridge University Press.

———. 2014. *Lectures on Anthropology*. Edited by Allen Wood and Robert Louden and translated by Robert Clewis, Robert Louden, G. Felicitas Munzel, and Allen Wood. Cambridge: Cambridge University Press.

Makkreel, Rudolf. 1990. *Imagination and Interpretation in Kant: The Hermeneutical Import of* The Critique of Judgment. Chicago: University of Chicago Press.

Matthews, Patricia M. 1996. Kant's Sublime: A Form of Pure Aesthetic Reflective Judgment. *Journal of Aesthetics and Art Criticism* 54 (2): 165–180.
McLaughlin, Peter. 1990. *Kant's Critique of Teleology in Biological Explanation: Antinomy and Teleology*. Lewiston: E. Mellen Press.
Zammito, John. 1992. *The Genesis of Kant's* Critique of Judgment. Chicago: University of Chicago Press.
Zuckert, Rachel. 2007. *Kant on Beauty and Biology: An Interpretation of* The Critique of Judgment. Cambridge: Cambridge University Press.

10

Equine Driving: Plato, Kant and Fichte on the Teamwork of the Mind

Günter Zöller

Everywhere I go I find a poet has been there before me.
—The quotation consistently has been attributed to Sigmund Freud, with no exact citation ever provided—and possible to give.

The contribution places the sustained recourse to the concept of drive in the accounts of practical subjectivity in J. G. Fichte into the historical and systematic context of Platonic and Kantian thinking about the psycho-politics of self-rule. Part 1 prepares Kant's, Reinhold's and Fichte's thinking about the unity and functionality of the mind in matters of desire and volition by turning to Plato's mytho-poetic comparison, in the *Phaedrus*, of the soul's set-up and manner of operation to a team of horses of opposed character that are driven by a seriously challenged charioteer. Part 2 first addresses Kant's account of the twofold, irrational and rational modes of practical subjectivity, along with the related conception of the

G. Zöller (✉)
University of Munich, Munich, Germany
e-mail: zoeller@lmu.de

M. Kisner, J. Noller (eds.), *The Concept of Drive in Classical German Philosophy*,
https://doi.org/10.1007/978-3-030-84160-7_10

will's twofold freedom as arbitrary election (*Willkür*) and legislative autonomy (*Wille*), and then traces Reinhold's and Fichte's appropriation of the concept of drive for detailing the dynamic structure and functionality of the mind's multiple and competing drives, including the "selfish" and "unselfish drive" in Reinhold and the "natural," "pure" and "mixed" or "ethical drive" in Fichte.

In drawing on Plato's equine analogy for the mind's regimen of self-rule, the contribution builds on the etymological identity of the German term *Trieb* and its English equivalent "drive." To be sure, the German *Trieb* does not share the extended semantic field of the English "drive," which includes the vehicular handling of animals, as in driving a team of horses. Moreover, in resorting to the Platonic modeling of the mind, the contribution pays close attention to the civico-political origin and horizon of much of the ethico-moral conceptuality employed by Fichte and his prime precursor, Kant, for capturing the complex and conflicted unity and operationality of the mind. In addition to providing the ancient poetico-philosophical context for modern accounts of practical subjectivity, the contribution aims at featuring the driven as well as driving nature of the modern subject in Kant, Reinhold and Fichte.

1 Ancient Hippo-Dramatics

> To begin with, our driver is in charge of a pair of horses; second, one of his horses is beautiful and good and from stock of the same sort, while the other is the opposite and has the opposite sort of bloodline. This means that chariot-driving in our case is inevitably a painfully difficult business. (Plato 1997, 542; Phaedrus 246b)

When reflecting on the diverse modes of rendering concepts sensible, Kant—in section 59 of the *Critique of the Power of Judgment*, which is charged with elucidating beauty as a symbol of morality (*Sittlichkeit*)—remarks that language is "full of [...] indirect presentations according to an analogy by means of which the expression does not contain the schema proper for the concept but merely a symbol for reflection" (Kant 1900, 5: 352; my own translation). The logical relation of analogy between two terms of signification, only one of which refers to a sensible object, while the other denotes an object of thought, introduces the linguistic devices of metaphor and trope as key conveyances for the thinking of that which cannot be the sensorily given, but which may prove to be an indispensable part of our intellectual and moral economy. According to Kant, our discourse, including philosophical discourse, is laden with such metaphorical, analogical or symbolic devices, which, each individually, may be subject to substitution and change but which, generically and collectively considered, are the indispensable tools of talk and thought.

A prime area for the Kantian symbolism of philosophical discourse is moral philosophy broadly conceived (*philosophia moralis*, moral science, *science morale*), encompassing ethics, law, politics and economics, and juxtaposed to natural philosophy equally broadly construed (*philosophia naturalis*). Kant himself gives the example of a hand mill analogically, symbolically or tropically rendering the modern absolutist state, only to immediately juxtapose to it the entirely alternatively constituted quasi-republican state ruled by "intrinsically popular laws" (*innere Volksgesetze*; Kant 1900, 5: 352; my own translation) likened to a living body or organism—a concept at the very heart of the third *Critique's* second part, on the teleological power of judgment, to which the section on symbolism forms the transition.

The Kantian device of symbolic representation in general and of philosophical symbolism in particular has a distinguished history. A paradigmatic instance of the latter is Plato's extended analogy between the soul and the city-state (*polis*) in the *Republic*. To be sure, in this case the symbolizing and the symbolized terms are not distinguished by being sensible and intelligible, respectively. Rather both domains, the psychological and the politological, are intellectual. But this is also the case with Kant's contrasting of the despotically and the republicanly ruled monarchical state—and with many other cases of philosophical metaphors that draw on a field of symbols already heavily informed by non-sensible terms and concepts.

The extended psycho-political analogy that underlies Plato's *Republic* has informed much subsequent thinking about the norms and forms of life in the city and the state. Particularly influential has been Plato's image of the quasi-political set-up of the soul on the basis of the three estates that make up Socrates' ideal city-state (Plato 1997, 1007f.; Republic 368d–369a). The focal point of the analogy has been the unity of the *psyche* and the unity of the *polis*, respectively, amid their analogous differentiations into distinct psychic and political powers, forces or capacities. In particular, there has been pervasive talk about the "parts" of the soul, in analogy to the city-state's political partition. Yet Plato himself is rather reticent about using overtly mereological language, instead preferring to talk about different kinds or species of the soul (*eide*) (Plato 1997, 1074f.; Republic 442c, 444b).

The simplistic talk about the soul having parts, to be found in a good many readings and interpretations of the *Republic*, is all the more troubling given Plato's (or rather Socrates') rejection of the soul's composite nature in the *Phaedo*, where the soul's immortality is at issue (Plato 1997, 91f.; Phaedo 106e). As Plato has Socrates argue in this dialogue, considered roughly contemporaneous with the *Republic* (both being middle Plato), what is composed of parts is subject to decomposition into its parts and thus in danger of losing its very unity and hence identity—the crucial moment for the soul's threatening disintegration being death and the ensuing decomposition of the human body.

A further feature that has caused puzzlement to readers of Plato past and present is the identification of the parts, the species or the functions that make up the soul as well as the city-state in the *Republic*. For here Plato famously inserts between the brainy and the appetitive regions of the body physical as well as the body politic (*logistikon*, *epithymetikon*) the irascible modality within the soul and the city-state ("spiritedness"; *thymoeides*) (Plato 1997, 1188; Republic 580–581a). Moreover, in its middle position, half way in between the head and the underbelly, chesty *thymos* is open to influence from either of its anatomical neighbors, lending impetus or drive to either the logical or the alogical regions, directions or orientations of the soul as well as the city-state.

To be sure, for Plato's Socrates, the ambiguous position of *thymos* is to be disambiguated in favor of the rule of the rational over the irrational by means of the thymotic energy element that is to serve as the auxiliary force for assuring the regimen of reason. Barring this right and just order of things psychic and political, the unity of the soul and the unity of the city-state are threatened with violent dissolution under the guise of civic strife, even civil war (*stasis*), and its psychic equivalent (Plato 1997, 1075; Republic 444a).

The exegetical *minutiae* involving parts (*mere*) and species (*eide*) notwithstanding, the *Republic* offers a strategically limited outlook on the relation between the three main elements of soul and city-state alike. Normatively considered, the logical element is to rule, the appetitive element is to be ruled and the thymotic middle is to assure reason's rule over the appetites. Given that in the *Republic* the politics of the soul closely follows city-politics, Plato's psychology and its embedded ethics also imitates the political outlook of the work, which seems distinctly illiberal when compared with, and contrasted to, modern conceptions of the state from Hobbes and Locke through Montesquieu and Rousseau to Kant and Humboldt, to name only the proto- and pre-liberals among the moderns.

One need not be an adherent of the view, championed by Karl Popper in the wake of twentieth-century totalitarianism on the extreme left as well as right, that Plato must be counted among the "enemies of the open society" (Popper 2013), to perceive the anachronism of Plato in the modern political world—the world of popular sovereignty, liberal democracy and free market economy, at least in the Western or Western-style parts of the world today. To be sure, there also have been attempts to retrieve, salvage or restore the standing of Plato's *Republic*, such as Kant's attempt in the *Critique of Pure Reason* to redefine the Platonic republic through the idea of civically compatible individual freedom in the state *qua* commonwealth (Kant 1998, A 316/B 372f.), or Hegel's attempt (later mimicked by Nietzsche) to read the *Republic* as an ideal-type depiction of Greek civico-political "ethical life" or "ethicality" (*Sittlichkeit*) resuscitated by Plato at the very movement of its demise brought about by the joint forces of Athenian direct democracy, anti-philosophical sophistics and post-Sophoclean, Euripidean experimental theater (Hegel 2003, 222f.).

But it is not only the ancient political and ethical values that seem to sit uneasily in the context of modern political and ethical life, together with its philosophical reflection in largely liberal political philosophy from J. St. Mill to Rawls and from B. Constant to Habermas. The very conceptuality employed by Plato some twenty-four centuries ago must seem alien to us moderns—given its key features that have since become marginal, such as *thymos*, or, inversely, given the concepts that are conspicuously absent in Plato but that have enjoyed an almost inflationary circulation in modern times, such as freedom or liberty.

In order to prepare the ground for possibly bridging the gap separating the ancients and the modern, but also separating the ancients among the moderns (conservatives) from the modern moderns (liberals), it might help to turn to another of Plato's dialogue for a differently designed image of the soul's set-up and function, one less politically fraught and more attuned to the dramatic outlook on psychic life cultivated by the moderns—as exemplified in dramatic literature from Shakespeare to Ibsen and in modern philosophical psychology from Spinoza to Freud. This alternative source for the psychodrama underlying ethics and politics is Plato's *Phaedrus*, another dialogue to be dated in the vicinity of the *Republic* and ostensibly concerned with matters of love, but also engaged in a mythically clad critique of writing (literality, *Schriftlichkeit*) as too rigid and generally mind-deadening a cultural technique.

Philosophically, a main preoccupation of the *Phaedrus* is with the proper place of "madness" (*mania*) in a human life form chiefly shaped by reason and self-discipline. The madness to be admitted, even required, under those conditions is qualified as "divine madness" (*theia mania*). In that vein, the *Phaedrus* portrays the proper proportion between the various factors and features that make up the soul in its optimal operation by means of a narratively detailed image inspired by classical Greek mythology and informed by an esteemed contemporary spectator sport (Plato 1997, 524–531; Phaedrus 246a–254e).

The myth involves a (winged) charioteer driving a chariot drawn by two (winged) horses, one of which is described as noble in appearance and character and of light color, the other as base and dark colored. The former horse pulls the chariot upward, while the latter drags it downward on a course suspended between the sky and the earth. It is the task of the driver to handle the horses, keep them on course and coordinate the often opposite pull exercised by the horses in the team. The imagery of the soul's chariot employed by Plato-Socrates draws on a Greek myth, mainly preserved through its much later literary rendition in book 2 of Ovid's *Metamorphoses*, about the chariot of the sun and the celestial

Phaethon traffic incident, which occurred when the chariot once was driven not by the sun god Helios himself, but by Helios' mortal son, Phaethon. Phaethon, who lacked training and experience, was unable to keep the chariot on course. When the sun chariot came too close to the earth and began to scorch it, Helios had to kill Phaethon with a lightning bolt in order to save the earth.

There is no precedent in the Phaethon myth, though, for the joining of oppositely natured horses within one team to be found in Plato's myth. But there is another myth involving a second and even a third mythological figure also named Phaethon (literally, "the shining one")—in one instance, Phaethon is the son of the goddess of dawn, Eos, the rose-fingered one in Homer's topical phrase, and in the second instance, Phaethon is the name of one of two horses of Eos, the other horse of the team being called *Lampos*, meaning "glitter." In an odd doubling of myths, Eos' chariot ascends the sky every morning before the sunrise performed by Helios in his own chariot. Compared to Helios' chariot, the one driven by Eos is more fitting as the background of the Platonic myth. For one, Eos' chariot is drawn by two horses, as opposed to the four horses pulling the sun god's chariot. Moreover, pictorial renditions tend to portray Eos in her two-horse chariot with wings on all three figures, the charioteer as well as the horses. Munich's State Antiquities Collection (*Staatliche Antikensammlung*) on the neo-classically designed *Königsplatz* has a fine specimen of the motive on a *krater* from Southern Italy (late fifth century BC).[1]

The other context that informs Plato's design of the soul's chariot is the actual deployment of the chariot in late Bronze Age Greek culture, especially in warfare, though more for transport to the battlefield than for actual vehicular fighting. Later, in the Greek military confrontation with Asia (first in the Persian Wars and then in Alexander's Asian campaigns), the Hellenic hoplites (heavily armored foot soldiers with long lances, acting in phalanx formation) proved superior to the enemies' deployment of war chariots. By that time, though, the chariot in Greek culture had become a piece of sports equipment, featured at Olympic and other

[1] For a digital reproduction, see https://de.m.wikipedia.org/wiki/Datei:Eos_chariot_430-420_BC_Staatliche_Antikensammlungen.jpg (retrieved 13 September 2020).

Panhellenic games. The aristocratic and bellicose pedigree of the chariot survived in chariot racing—it being the most expensive sports discipline to engage in. Typically the chariot was not even driven by the owner of the chariot and its team of horses but by a hired athlete. The price, if any, still went to the team's owner, as in today's world of equine contests.

The relevance and revelatory potential of the Greek cultural background in chariot driving for Plato's mythical chariot of the soul becomes strikingly evident in another antiquity from fifth-century BC Greece, this one not preserved in Munich (the "Athens on the Isar," so named after the tributary of the Danube that runs through it) but where it belongs, given its original function as a dedicatory statue in the sanctuary of Delphi. It is the Charioteer of Delphi, one of the few ancient Greek bronzes still originally preserved, rather than known through later, Roman copies.[2] The statue is life size if not slightly larger (1.80 meters), portraying the fully clothed youthful charioteer as he presents himself on the chariot with the horses, now lost, to the spectators of the race he just won.

For present purposes, what matters most about the Delphic charioteer is the figure's restraint in the very moment of victory and glory—a posture of self-constraint and self-control aided, in fact conveyed, by the objectivist style of the statue, which dates from 480 BC, the year of the second Persian invasion, and thus belongs to the brief transitional period between the late archaic and the early classical style, a strikingly suspended style also represented, grandiosely and gloriously, by the preserved pediment sculptures from the East side of the Aphaia temple on the island of Aegina opposite Athens. Together with the remaining, more clearly archaic sculptures from the West pediment of that temple, the remains—the Aegina Marbles, as opposed to London's Elgin Marbles—are housed in Munich's other antiquities collection, the *Glyptothek*, also forming part of the neo-classical ensemble of the *Königsplatz*.[3]

---

[2] For a digital reproduction, see https://en.wikipedia.org/wiki/Charioteer_of_Delphi#/media/File:AurigaDelfi.jpg (retrieved 13 September 2020).

[3] For a digital reproduction, see https://commons.wikimedia.org/wiki/File:West_pediment_-_Temple_of_Aphaia_in_Egina_-_Glyptothek_-_Munich_-_Germany_2017_(2).jpg (retrieved 13 September 2020).

The image of control, foremost including self-control, that is conveyed by early classical sculpture in general and by the examples from Delphi and Aegina in particular is apt to visualize the demeanor of the charioteer in Plato's mythic tale, who, while controlling both of his horses, also and most importantly controls himself. The myth, as fabricated by Plato for philosophical purposes, is not about letting the good horse run its course, while reigning in the bad one. Neither is the myth about choosing between horses and making either the right or the wrong choice about the winning horse, so to speak. Rather it is about exercising control and remaining in control over the entire chariot set-up and about knowing where to go in the first place. The myth thus is about the precarious position of the charioteer, who finds himself, as it were, half way between the earth and the skies, between the world of sense and the world of spirit, and who needs all the control and self-control he is able to muster to steer his course. But the myth is also, in line with the erotic subtext of the entire dialogue, about self-abandonment in madness, not in madness of all sorts but in madness of the right sort ("divine madness"), and about the control of one's self to be retained even and especially in such madness according to method.

The congruence in character between Plato's myth of the charioteer and early fifth-century BC Greek sculpture is no accident. The discrepancy in chronology between Plato's early fourth-century writing and art from an entire century earlier reflects the pointed anachronism of Plato's philosophy and politics. Not for Plato the liberal individualism of Periclean Athens, as perceptively portrayed in Thucydides' account of Athens at the beginning of the Peloponnesian War (Thucydides 2009, 90–96). With his aristocratic leanings, his lakonophilia (love of things Spartan) and his grudge, not to say resentment, against mass-democratic Athen on account of the populist trial, condemnation and execution of his teacher Socrates, Plato must have seemed a conservative revolutionary in post-war Athens—not to say, in Attic Weimar.

Still Plato's nostalgia for earlier values and views, for what Hegel came to call "Greek ethical life" (*griechische Sittlichkeit*; Hegel 2003, 223), brings out a politico-philosophical alternative to Athenian (proto-)liberalism and a kind of freedom that is unlike the liberty of free choice—a praeter-liberal form of freedom cherished by political and moral philosophers from Plato and Aristotle to Kant and Fichte. To Plato and his contemporaries (politicians, philosophers and citizens alike), freedom or liberty—the Greek term for both of them being *eleutheria*—could mean many things. The base meaning derived from the free master's absolute dominion over the unfree slave (*despotes* and *doulos*, respectively). Politically speaking, to be free was to rule, rather than to be ruled, as in the fragile freedom of the Greeks from Persian rule. When applied to an individual Greek *polis*, freedom also meant the freedom from despotic rule by a local strong man ("tyrant"). Transposed from the political sphere to that of the individual human being, freedom qua self-rule came to mean ruling over one's passions, appetites and desires, rather than being ruled by them. In analogy to the twofold political freedom from foreign rulers and from domestic tyrants, the psycho-ethical freedom of self-rule was not only opposed to the unfreedom of being ruled by someone else but also to the pseudo-freedom of being ruled by one's own passions, appetites and desires in an illusionary scenario of secretly being ruled while seemingly ruling oneself. Such is the case of the "tyrannical human being" in the psycho-political pathology provided by Plato's *Republic* (Plato 1997, 1180; Republic 571a).

2 Modern Psycho-politics

> He used to call his upper and lower forces of the soul the Upper and Lower House, and very often the former passed a bill that the latter rejected. (Lichtenberg 1967–74, 1: 65; Sudelbücher, B 67)

The political background of the ancient Greek linkage of freedom to self-control finds a modified continuation in modern conceptions of freedom that are political in origin and psychological in application. With the advent of the modern sovereign territorial state in post-feudal Europe, as theorized by Jean Bodin and Thomas Hobbes, political rule, regardless of its outwardly monarchical, aristocratic or democratic form, becomes tied to the care for a body politic constituted by sovereign and subjects alike ("commonwealth"; Hobbes 1998, 114; *république*; Bodin 1992, 90 [translated as "commonwealth"]). The newly envisioned and provided freedom of the citizen—"civil liberty of the people" (Montesquieu 1989, 167)—consists in entitlements ("liberties") to all things not expressly prohibited by law ("silence of the law"; Hobbes 1998, 146), eventually formalized in the conception of praeter-positive legal claims ("Law of Nature"; Locke 2003, 274).

In a related development, to be observed in some European countries, the citizenry collectively considered acquires freedom in a specifically political, as opposed to merely civil, sense—"political freedom" (*liberté politique*; Montesquieu 1989, 157)—by forming part of the legislation of the land. Foremost in England, freedom thus assumes the twofold meaning of being negatively free, or "at liberty," to do as one pleases, and positively free in giving oneself laws through representatives of one's choosing or, at least, from among one's constituency. Historically speaking, parliamentary representation, however limited and venial, provides a modern successor of sorts to direct-democratic self-rule in ancient Greece and to pseudo-democratic client politics in Republican Rome—one deemed superior in principle and fact over the ancient systems though (Zöller 2021).

To be sure, the modern system of parliamentary representation, as advanced in England and practiced in Continental republican regimes, for example, in the Netherlands ("States General"), typically is limited to the legislative branch of government and leaves the executive in the hands of a hereditary monarch or a self-selected aristocracy. Technically speaking, the newly acquired, or at least desired, freedom is not self-rule but self-legislation—autonomy in the original, legislative sense of the term. Only the political upheavals ("revolutions") at the end of the eighteenth century, first in British colonial North America, then in bankrupt absolutist France, bring, with varying success and duration, popular sovereignty and democratic government to an extended territory far exceeding the narrow confines of an ancient Greek city-state.

The psychological fallout of modern politics and modern political philosophy is foremost visible in Kant, who builds his entire mature moral philosophy on the "principle of autonomy" (Kant 1900, 4: 433). To be sure, the ethical lawgiving propagated by Kant is not some psychic parliament's majoritarian decision process on norms of conduct to be adopted and followed by all. Rather than involving election and selection, the legislation procedure embraced by Kant involves reason providing the moral law and the moral agents giving the law unto themselves. Ethical autonomy in Kant is not a case of personal self-determination, as in current conceptions of individual, social, civic or political autonomy. Instead, the *nomos* in "autonomy" is reason's own law, and the *autos* in "autonomy" is reason itself, which both issues and determines the law.

The dual, formal as well as contentual autonomy of reason in Kant's ethics brings with it the extraordinary character of the moral law in Kant. Far from involving a material supreme end, as in ancient eudaimonist ethics under the teleology of some highest good, the basic law of Kantian morality, featured in *Foundation of the Metaphysics of Morals* (1785), is nothing but the very form of "lawfulness" (*Gesetzmäßigkeit*) as such (Kant 1900, 4: 402). For Kant, ethical action, or rather the volition that is to underlie it, consists in willing under the law's minimal, merely formal condition of "universality" (*Allgemeinheit*) (Kant 1900, 4: 421). And this for no other reason than that willing under any different form would involve heteronomous factors ("inclinations") obstructing the pure rationality of willing and impeding the strict autonomy of reason.

Kant freely concedes that ordinary willing occurs under essentially heteronomous conditions, involving self-interest as well as self-deception about the factual involvement of self-interest. Yet when it comes to assessing the moral character ("morality") of willing, all heteronomous factors are to be disregarded and the universal form of willing is all that matters. In his late (quasi-)material ethics under the guise of a Doctrine of Virtue (*Tugendlehre*), in Part Two of the *Metaphysics of Morals* (1797), Kant even has the previously established "autonomy" (*Autonomie*) of ethical willing enhanced by the "autocracy" (*Autokratie*) of ethical reason in the determination of the will (Kant 1900, 6: 383). The autonomy of reason consists in the giving of reason's moral law and is modeled on the political power of legislation. By contrast, the autocracy of reason resides in the habituated capability to enact the moral law of reason against the countervailing influence of the inclinations and is modeled on the executive political power. For Kant, the ancient acquired character disposition of "excellence" (*arete*) rationally reduces to "virtue" (*Tugend*) as the studiously cultivated propensity to act according to the moral law and because it is the moral law, rather than for other, selfish reasons, or even contrary to the moral law altogether (Kant 1900, 6: 383f.).

The difference between Plato's essentially ancient conception of the soul's rational rule over itself and Kant's decidedly modern notion of autonomous (and autocratical) rational willing notwithstanding, Plato's and Kant's accounts agree on an ethos of self-control and self-discipline that is to permeate a human life worthy to be called "free." But for Kant mere self-control, or even self-control by rational means, is not sufficient for ensuring ethically meaningful freedom. For there is, on Kant's view of the matter, the not uncommon case of the self's very own sensibility coercing its rationality into service, thus instrumentalizing reason, which amounts to a kind of "self-coercion" (*Selbstzwang*), though one that is essentially unfree (Kant 1900, 4: 394).

According to Kant, only a self-coercion that involves reason and reason alone constraining sensibility, independent of any sensory influence, can be considered a case of genuine, "self-coercion" involving "inner freedom" (Kant 1900, 6: 395). For Kant, ethical self-coercion and free self-coercion coincide, and vice versa. By contrast, to Plato's essentially ancient sensibility, the very notion of self-control excludes a ruling role of the senses, desires or appetites, given that human beings are by nature—by their own nature—either such that they are to rule or such that they are to be ruled, and hence are not subject to choosing and changing in that regard. To put the matter in terms of the social relationship serving as a foil to ancient Greek thinking about freedom and unfreedom: the senses are the "natural slave" (Aristotle 1998, 7–9; Politics 1254a18–1255a1) to reason.

To modern sensibility, the implicit enslavement of the senses in the ancients not only fails to recognize the genuine contribution of sensibility to the economy of the mind, as ascertained in an academic manner by G. A. Baumgarten's emancipation of aesthetics from logic. For some modern thinkers, such as Hume, the ancient outlook also errs in failing to recognize the essentially enslaved position of reason with regard to the passions ("the slave of the passions"; Hume 2007, 1: 266). To be sure, Kant does not share the sentimentalist and misological assessment of reason's solely subservient role. On the contrary, on Kant's considered account, presented in the *Critique of Practical Reason* (1788), reason is not only able to motivate willing but to sufficiently motivate it all by itself, as disclosed in the famous, or rather infamous, "fact of reason" (*Faktum der Vernunft*), to be understood as the immediately certain and practically efficient awareness of one's unconditional moral obligation (Kant 1900, 5: 31).

To be sure, previously Kant, too, had considered reason to be able to provide at most the criterion for assessing the morality of willing and acting (*principium diiudicationis*), while he regarded the force underlying the actual execution of moral willing and acting (*principium executionis*) to be a feeling *sui generis*, a moral sense (*moralisches Gefühl, sensus moralis*) (Kant 2004, 22 and 55f.). By the mid-1780s though, as documented in *Foundations of the Metaphysics of Morals*, Kant came to entrust reason—more precisely practical, will-determining reason—with sufficient pull, or push, to generate an entire volition (or action), this constituting the rare, in fact the sole case of pure reason being practical all by itself ("pure practical reason," also "practical pure reason") (Kant 1900, 4: 389).

The sufficiently motivating potential of (practical) reason claimed by the critical Kant has seemed to many a cheap solution in the manner of a *deus ex machina*—a divine device miraculously dropped onto the theater of the modern mind, coming out of nowhere and providing a dramatic solution that may entertain but cannot satisfy. To Kant, though, the introduction of purely practical reason must have seemed the only option in a scenario that excluded the morality of feeling, embraced by British, chiefly Scottish moralists, and the perfectionist moral teleology of an ethics of essentially theoretical, merely cognitive reason, espoused by Continental rationalists.

Yet Kant seems to have overlooked the Platonic solution to the ethical ineffectiveness threatening mere reason, which consists in taking recourse to a force or source, *thymos*, at once distinct from rational thought and from the irrational appetites or desires—a force that could be made to aid reason in the latter's effort to subdue the irrational, much like a general employs an army to fight the enemy. But as far as ancient leanings go, Kant was neither a *connaisseur* nor a follower of Plato's aristocratically inclined moral psychology and ethics.

To the extent that his ethical formalist rationalism permitted, or even required, Kant was a neo-Stoic or rather a neo-neo-Stoic, influenced by late-Roman, imperial-age and Latin-language neo-Stoicism, such as Seneca's, who teaches and practices the mastery of any and all passions, including such potent motivators of private and political action as "anger" (*ira*), in order to establish and maintain a serene mind-set (*tranquilitas animi*). To Kant's twice-removed Stoic sensibilities, Platonic *thymos*, with its linkage of virtue to virility (as etymologically conveyed by the Latin term for "virtue," *virtus*), would have appeared a most unreliable ally prone to turn against reason and establish its own martial and manly counter-regimen.

It was left to one of Kant's closest followers to provide a variation on the Platonic theme of the soul's two-horse chariot, though informed not by an ancient image from warfare and competitive sports but by a modern notion for the driving force behind the various phenomena of life and living beings. The term "drive" (*Trieb*), introduced by K. L. Reinhold into philosophical psychology, had had a recent and short history in nascent biology ("natural history"), where it served to label the dispositions for the shaping of a living being with regard to the latter's parts ("organs") and whole ("organism"). The chief advocate of the discourse of drive had been the Göttingen physician and physiologist, Johann Friedrich Blumenbach (1752–1840), who had sought to trace the origin and growth of living organisms to an underlying "formative drive" (*Bildungstrieb*; Blumenbach 1781) that was supposed to be less a newly discovered manifest feature of animal life than an explanatory hypothesis of its overall organization and vital dynamics (Witte 2019).

Already Kant had approvingly accepted Blumenbach's interpretive device of the formative drive in the philosophy of biology that forms the core of Part Two of the *Critique of the Power of Judgment* (1790) (Kant 1900, 5: 424). But it was Reinhold who, in the second volume of his *Letters on the Kantian Philosophy* (1792), imported the notion of drive from natural philosophy into moral philosophy, the latter widely conceived as including the mind as well as morals (Reinhold 1792). Reinhold's *Letters*, especially the second installment that shifts the focus from the first *Critique* to the second *Critique*, combine introductory and explanatory aims with attempts at revising, if not the substance, then at least the presentational form (*Darstellung*) of Kant's academically geared philosophy in the interest of greater popularity and wider acceptability.

In *Letters II*, especially in the Seventh and Eighth Letter, Reinhold substitutes the Kantian dual disposition of the "faculty of desire" (*Begehrungsvermögen*), which had divided the latter into a higher, reason-ruled and a lower, purely appetitive register, with a duality of drives each identified through its relation to the self. While both drives, distinguished by Reinhold in *lieu* of the higher and lower forms of desire, refer to the self whose drives they are and whom they are to propel into action (or move to inaction), the two essentially self-serving drives are differentiated as "selfish" (*eigennützig*) and "unselfish" (*uneigennützig*), respectively (Reinhold 1792, 220–261 and 262–308).

On Reinhold's view, the self-regard that is definitional of a living being's outlook and outreach manifests itself in two forms: as selfish self-regard, narrowly focused on egoistic aims and ends, and as unselfish self-regard, generously geared toward a more encompassing conception of self, akin to the ancient Greek virtue of "magnanimity" (*megalopsychia*)—an Aristotelian reference not to be found in Reinhold himself though. Faced with the two kinds of desire, reconfigured as two opposed drives, the Reinholdian agent so suspended between opposite drives is to choose which of them to satisfy, with the process of deciding among them constituting an exercise of "free arbitrary choice" (*freie Willkür*).

Kant was neither pleased nor convinced by Reinhold's alleged emendation, which he regarded as making things worse rather than better, a reversal of outcome for which the German language knows the term, *Verschlimmbesserung*, meaning "worsening improvement." In particular, in the *Metaphysics of Morals*, Kant objected to the parity maintained by Reinhold between the selfish and the unselfish drive with regard to a choice reduced from election guided by reason to the exercise of arbitrary decision. Moreover, Kant denied that the alleged faculty, capacity or ability (*Vermögen*) to follow the selfish drive and thereby act against the moral law, as adduced by Reinhold, could be considered a positive potentiality at all, when in fact it had to be regarded as the entirely negative *in*ability, *in*capacity or *non*-faculty to follow the moral law (*Unvermögen*) (Kant 1900, 6: 227). In the same vein, for Kant the alleged freedom to let oneself be driven by the selfish drive constituted a failure of freedom and a lack of truly free, autonomous willing.

Plato would have shared Kant's qualms when seeing his horse-drawn chariot replaced by a duo of opposed drives and the rational charioteer seeking to gain and maintain control of the entire set-up being exchanged for an arbitrarily choosing drive elector, as though the drives were genuinely alternative modes for the vehicle's composite motion. In particular, Plato would have objected to the very notion that the alternative between the selfish and the unselfish driving force was really open, and the choice between the two of them arbitrarily free and equal. From a Platonic viewpoint, essentially shared by Kant, Reinhold's free choice among equipollent drives confuses the noble freedom consisting in recognizing and following reason and the base freedom of doing what one pleases, regardless of the actual grounds and consequences of such pseudo-liberal license.

While ostensibly moving away from Kant (and Plato) farther yet than Reinhold had done, J. G. Fichte continues the discourse of drives introduced into post-Kantian philosophy by retaining, rather than disregarding, the Kantian conception of positive, ethically requisite freedom (Jacobs 1967). To begin with, Fichte replaces the Reinholdian dualism of the selfish and the unselfish drive with the distinction between the natural and the pure drive, a differentiation meant to better capture Kant's distinction between the lower and the higher faculty of desire. In particular, Fichte conceives of the pure drive as the inherent striving after "freedom for its own sake" (*Freiheit um der Freiheit willen*) (Fichte 1965, 4: 139; my own translation), which thus forms the ultimate as well as the highest goal of human endeavors. In the context of his post-Kantian transcendental philosophy (*Wissenschaftslehre*), Fichte understands freedom as independence from anything alien and other (*Selbständigkeit*) in the interest of unalloyed, purely rational self-identity or I-hood (*Ichheit*) (Fichte 1965, 4: 14).

By contrast, the natural drive for Fichte involves the striving for material ends of all kinds, along with the means for achieving them. The particular ends and their suitable means are given by the nature of things—by the naturally given disposition of the individual human being in question, but also by the natural circumstances of life at hand. On Fichte's outlook, the two drives are not so much opposed to each other, as that they represent the different sides of the human being, which is, at once, finite and infinite, sensible and intelligible, beastly and divine on Fichte's proto-existentialist theo-anthropology of striving subjectivity (Fichte 1965, 4: 130). Accordingly, the overall strategy of Fichte's reconceptualization of Kantian practical psychology is integrationist and reconciliatory, to the point of envisioning the fusing of the artificially abstracted pure drive and the equally abstract natural drive into the concrete, reconciled "ethical" drive.

On Fichte's post-Kantian construal, the ethical drive is to take its ultimate end—an ultimately infinite end—from the pure drive, which aims at nothing but absolute independence ("freedom"), and its intermediary, entirely finite ends from the natural drive, which represents the demands of the limited, materially manifest aspects of human existence. For Fichte, the fusion of the material natural drive and the formal pure drive results in the serialization of human agency: in each moment, the human agent is to act in such a way that the limited end pursued under the given circumstances is to contribute to an entire consistent series the infinitely remote terminal point of which is absolute freedom (Fichte 1965, 4: 131). The details of Fichte's vision of the moral order approached in so many ways by multiple parallel series involving distinct individual agents remain largely unclear, even unaddressed. Fichte seems to assume a pre-established harmony of sorts between the parallel series in their joint effort to establish complete independence at the price of utter ethical identity between naturally diverse agents. For Fichte, the ethics of absolute freedom culminates in the prediction, more so than the prescription, that ultimately every ethical being will *act* the same because all ethical beings *are* the same.

To be sure, Fichte's post-individualist scenario of absolute ethical identity remains an infinitely removed end, ever to be approached and never to be achieved. But it reflects a politico-philosophical mind-set that equates total freedom with total identity—identity with oneself as well as with (all) others—at the expense of difference and diversity at the individual and social level. Fichte's ethics of the mixed drive has substituted Kant's ethical criterion of the universality of the moral maxim with the (pseudo-)ethical requirement of a comportment common to all. Kantian (proto-)liberalism has been exchanged for Fichtean (proto-)socialism (Zöller 2020). In a related development, Fichte's ethics eschews recourse to the rationally structured device of the categorical imperative, instead basing the moral assessment of self and others on the allegedly unerring voice of conscience (*Gewissen*) (Fichte 1965, 4: 147 and 194).

With the Kantian procedural rationality requirement removed from Fichte's ethics, the orienting as well as motivating force of action resides entirely in the drives—in the overall orientation provided by the pure drive and in the impetus to particular, serially suited actions promoted by the natural drive. All that is left to the driver—Plato's former charioteer in charge—is to carefully coordinate the short-term and the long-term drives, which, first by themselves and then jointly, propel the chariot, with little need for steering or guiding on the part of the driver, who is more a passenger than a charioteer in Fichte's identity-focused psycho-mobile.

The freedom that comes with Fichte's post-Kantian retake on the chariot of the soul is at once major and minor. It consists, materially, in the sought-after but actually unattainable absolute freedom of utter independence and complete self-identity ("material freedom") (Fichte 1965, 4: 139) and, formally, in the freedom from necessitation through any drive that has not been found to fit into the overall serial scenario of ethical progress ("formal freedom") (Fichte 1965, 4: 135). With the Kantian moral law replaced by the postulate of sheer self-identity and Kantian pure practical reason replaced by the voice of conscience, Kantian autonomous and autocratous freedom—an initiative freedom according to a self-given law—also receives a radical replacement in Fichte; it reduces to the somewhat ceremonial function of consenting or dissenting to what the composite character of the mixed drive pushes for. Based on Kant's derogatory remark about the pseudo-freedom of a wound-up mechanical turnspit (*Freiheit eines Bratenwenders*) (Kant 1900, 5: 97), one could call this Fichtean formal freedom the "freedom of the rubber stamp."

Looking back across Fichte and Reinhold to Plato, it seems that the ancient function of the charioteer has become obsolete. The modern soul appears largely moved by drives of various kinds, with little else to do for the remaining driver on the chariot than to react to competing, conflicting or coalescing drives. It is entirely in line with this takeover of the chariot by the horses, when Freud declares, a century after Reinhold and Fichte, that the rationally free self ("the I") is not the master in its own house (Freud 1917, 7). Kant would have agreed with this sober assessment, given the realist, not to say pessimist anthropology that forms the reverse of his idealist ethics of rationally free autonomy (Zöller 2011). The sublime thought of reason being practical all by itself and hence sufficient to determine the will seems to have turned into a tall tale not unlike Plato's uplifting myth about the charioteer of the soul.

References

Aristotle. 1998. *Politics*. Trans. C. D. C. Reeve. Indianapolis and Cambridge: Hackett. Cited after the standard "Bekker" pagination indicated in most modern editions and translations.

Blumenbach, Johann Friedrich. 1781. *Über den Bildungstrieb und das Zeugungsgeschäfte*. Göttingen: Dieterich.

Bodin, Jean. 1992. *On Sovereignty: Four Chapters from The Six Books of the Commonwealth*. Ed. and Trans. Julian H. Franklin. Cambridge and New York: Cambridge University Press.

Fichte, Johann Gottlieb. 1965. Das System der Sittenlehre. In Johann Gottlieb Fichte, *Sämtliche Werke*, ed. Immanuel Hermann Fichte, vol. 4, 1–365. Berlin: De Gruyter. The pagination of this edition is indicated in most modern editions and translations of the work.

Freud, Sigmund. 1917. Eine Schwierigkeit der Psychoanalyse. *Imago: Zeitschrift für Anwendung der Psychoanalyse auf die Geisteswissenschaften* 5 (1): 1–7.

Hegel, Georg Wilhelm Friedrich. 2003. *Elements of the Philosophy of Right*. Ed. Allen W. Wood and trans. H.B. Nisbet. Cambridge and New York: Cambridge University Press.

Hobbes, Thomas. 1998. *Leviathan*. Ed. J.C.A. Gaskin. Oxford and New York. Oxford University Press.

Hume, David. 2007. *A Treatise of Human Nature. A Critical Edition*. Ed. David Fate Norton and Mary J. Norton, vol. 2. Oxford: Clarendon Press.

Jacobs, Wilhelm G. 1967. *Trieb als sittliches Phänomen: Eine Untersuchung zur Grundlegung der Philosophie nach Kant und Fichte*. Bonn: Bouvier.

Kant, Immanuel. 1900. *Gesammelte Schriften*. Ed. Prussian Academy of Sciences and its successors. Berlin, later Berlin and Boston: Reimer, later De Gruyter. The pagination of this edition is indicated in most modern editions and translations of Kant's works.

———. 1998. *Critique of Pure Reason*. Ed. and Trans. Paul Guyer and Allen W. Wood. Cambridge: Cambridge University Press.

———. 2004. *Vorlesungen zur Moralphilosophie*. Ed. Werner Stark. Berlin and New York: De Gruyter.

Lichtenberg, Georg Christoph. 1967–1974. *Schriften und Briefe*. Ed. Wolfgang Promies, vol. 6. Munich: Hanser.

Locke, John. 2003. *Two Treatises of Government*. Ed. Peter Laslett. Cambridge, and New York: Cambridge University Press.

Montesquieu. 1989. *The Spirit of the Laws*. Ed. Anne M. Cohler, Basia Carolyn Miller and Harold Samuel Stone. Cambridge and New York: Cambridge University Press.

Plato. 1997. *Complete Works*. Ed. John Cooper. Indianapolis and Cambridge: Hackett. Cited after the standard "Stephanus" pagination indicated in most modern editions and translations.

Popper, Karl. 2013. *The Open Society and Its Enemies*. Princeton: Princeton University Press.

Reinhold, Carl Leonhard. 1792. *Briefe über die Kantische Philosophie*. Vol. 2. Leipzig: Göschen.

Thucydides. 2009. *The Peloponnesian War*. Trans. Martin Hammond. Introduction and Notes by P. J. Rhodes. Oxford and New York: Oxford University Press.

Witte, Egbert. 2019. *Bildungstrieb: Zur Karriere eines Konzepts zwischen 1780 und 1830*. Hildesheim and New York: Olms.

Zöller, Günter. 2011. Between Rousseau and Freud. Kant on Cultural Uneasiness. In *Rethinking Kant*, ed. Oliver Thorndike, vol. 3, 52–77. Newcastle upon Tyne: Cambridge Scholars Publishing.

———. 2020. Freedom, Right, and Law: Fichte's Late Political Philosophy. In *The Bloomsbury Handbook to Fichte*, ed. Marina F. Bykova, 261–275. London, New York, Oxford, New Delhi, and Sydney: Bloomsbury Academic.

———. 2021. 'Participation of the People Through Its Deputies.' Montesquieu, Kant and Hegel on German Freedom. *Graduate Faculty Philosophy Journal (New School for Social Research)* 42 (1) (forthcoming).

11

"The Drive to Be an I Is at the Same Time the Drive to Think and to Feel": Hardenberg/Novalis on Drives, Faculties, and Powers

Violetta L. Waibel

When we speak of drives, we speak of intentionality, of purposiveness, of ends, even of finite ends. So, with drives, whole cosmologies, whole philosophies, the philosophical universe from Plato, to Kant, to Fichte, to Freud are present. When we think about drives from the perspective of Friedrich von Hardenberg, also known as Novalis, in what is called *Fichte Studies* (1795/1796) we encounter a microcosm of elementary functions of drives. When Hardenberg speaks of drives, he nearly always speaks of mental actions and not of desires and the faculty of desires. In *Fichte Study* 325 he states: "Drive is activity of a certain sort—There are only drives in the representing and intuiting subjects as such—Passion is suffering of a certain sort" (Hardenberg 2003 [1795/1796], FS 325, 121).[1]

[1] Here and further, in the case of references to classics such as Hardenberg/Novalis or Fichte, the years of first publication or creation and, in the case of the *Fichte Studies*, the initials FS with the number of the study are given in order to make it easier for readers to find the textual references who do not have the relevant English translation at hand or who wish to read the German text.

V. L. Waibel (✉)
Institut für Philosophie der Universität Wien, Wien, Austria
e-mail: violetta.waibel@univie.ac.at

M. Kisner, J. Noller (eds.), *The Concept of Drive in Classical German Philosophy*,
https://doi.org/10.1007/978-3-030-84160-7_11

One important function of drives is that of an original act—*Urhandlung*. Closely connected to this is another important function, the *ordo inversus* of feeling and thinking that offers a theory on the surfacing of conscious contents from the unconscious. The unconscious that plays an important role in Hardenberg's struggle for a theory of knowledge is not an equivalent of Sigmund Freud's dynamic unconscious (Freud 2003 [1915]; Waibel 2005). In Hardenberg's *Monologue* of 1798/1799 we encounter a drive of language (*Sprachtrieb*) embedded in an ironical play of contradicting the normal communicative role of language. Here we encounter a use of negativity that reminds us of the *ordo inversus*. The term *drive* is one of the pivotal, systematically fundamental terms, which Hardenberg uses in his *Fichte Studies*. Systematically closely linked to drive is the question of what the relationship between the pure or absolute I and the empirical I looks like, and how the body's instincts enter into consciousness and establish an interrelation with thinking, and generally, with spirit. In this context Hardenberg coined the term *ordo inversus,* a theorem, which is indeed systematically affiliated to Johann Gottlieb Fichte's "Wechselbestimmung".

Let me first give a general overview of the *Fichte Studies* as well as of the writer and poet—as well as philosophical thinker—who was known as Novalis, but whom I prefer to call by his actual name—Friedrich von Hardenberg.

The *Fichte Studies* are an unpublished collection of manuscripts, the current edition of which was first compiled in 1965. The earliest and first editions consisted of pieces of the collection selected at the discretion of the editors. In 1930 the collection was auctioned off, taken abroad and no longer accessible to the scientific world until it resurfaced at auction in 1960 and was acquired by the Freie Hochstift in Frankfurt am Main. Hans-Joachim Mähl analysed papers types as well as the handwriting, and dated and arranged the collection on the basis of his findings. Finally, he published the present edition in 1965. Moreover, he divided the manuscripts written in the years 1795/1796 into six groups according to the working cycles he had identified, and gave all of it the name *Fichte Studies*

I owe many thanks to Manja Kisner for the careful revision of this contribution.

in reference to the obvious terminology taken from Johann Gottlieb Fichte's early *Wissenschaftslehre*.

In rearranging the manuscript, Mähl distinguished not only six groups of manuscripts but also 667 individual studies and defined the following periods (Mähl 1965):

1st Group: Studies 1–210: Fall to early winter, 1795
2nd Group: Studies 211–287: Winter 1795 to February 1796
3rd Group: Studies 288–372: February–March 1796
4th Group: Studies 373–552: March to early summer 1796
5th Group: Studies 553–568: Summer 1796
6th Group: Studies 569–667: Summer to fall 1796

I would prefer the edition to be called *Kant Fichte Studies*. Although Hardenberg obviously deals with Fichte's early *Foundation of the Entire Wissenschaftslehre* (1794/1795), he also clearly shows in the collection that he has profound and detailed familiarity with especially Kant's first *Critique*, that is, *The Critique of Pure Reason*, but also with various other parts of the *Critique*.

Although Hardenberg studied Fichte's *Foundation of the Entire Wissenschaftslehre* and Kant's *Critique of Pure Reason* very independently, he acquired his knowledge mainly by studying philosophy at the Universities of Jena and Leipzig. At times, he even spent more time studying philosophy than on studying to become a lawyer, which he intended to become. He enrolled at the University of Jena on October 23, 1790, and a year later at the University of Leipzig on October 24, 1791. In March 1793 he transferred to Wittenberg, where he put less emphasis on his philosophy studies and instead devoted himself to his law studies. On June 14, 1794, he graduated by passing the final exam (Hardenberg, NS V, 369–377; Eckhardt 1987, 21–46).

The *Fichte Studies* are Hardenberg's attempts at philosophical self-conception. He takes up individual themes of the Fichtean *Foundation of the Entire Wissenschaftslehre*, analysing them in part with his more or less profound knowledge of Kantian theorems, and attempts to reconstruct them more or less independently. The *Fichte Studies* are an inhomogeneous collection of notes, reflections, philosophical sketches of his own thoughts and interesting, sometimes even surprising ideas (*Einfälle*).

In this chapter, I will present some key passages of Hardenberg's that speak of drives and then move on to his remarkable reflections on the interrelation between feeling and thinking, which he also discussed as *ordo inversus*.

It is worth mentioning that Kant does not speak of drives in his moral philosophy, but of incentives (*Triebfedern*). Hardenberg may have studied drive theories in the writings of Karl Leonhard Reinhold, who he studied with in Jena, especially his *Versuch einer neuen Theorie des menschlichen Vorstellungsvermögens (Attempt at a New Theory of the Human Imagination).* Ernst Platner, who published *Philosophische Aphorismen* (*Philosophical Aphorisms*) in two parts in 1776 and 1782 (again in 1793), taught in Leipzig. Fichte worked out a theory of drives in the practical part of the *Foundation of the Entire Wissenschaftslehre* of 1794/1795. Fichte continues his theory of drives in his *System of Ethics*, published in 1798, where he elaborates a much more complex and differentiated theory of drives. Important is his idea of a mixed drive, composed of a pure drive of thoughts and the sensible aspects of nature and volition (*Begehren*) (Fichte 2005 [1798], 145). Hardenberg expressed a comparable idea in the *Fichte Studies* a few years earlier, focusing on mental activities and the relationship between the pure or absolute I and the empirical I without considering desires in detail.

1 Hardenberg's Thoughts on Drives

In *Fichte Study* 32, Hardenberg asks, with obvious reference to Fichte: "How does the absolute I become an empirical I?" (Hardenberg 2003 [1795/1796], FS 32, 24). His answer follows immediately and begins with the words I have chosen for the title of this paper: "The drive to be an I is at the same time the drive to think and to feel. Both manifest themselves in matter and in form—against one another. Now reflection has obtained *its determinate manner of acting* (its determinate material), and feeling has obtained its determinate form. *To* each other, both appear to be independent. The need of their drives is satisfied—they do not know how—that is, it does not lie within their spheres" (Hardenberg 2003 [1795/1796], FS 32, 24).

What is special about Hardenberg's approach, in my opinion, is that he attaches immense significance to the interrelation between thinking and feeling as well as to that between reflecting and feeling. He touches on this very subject in the same paragraph when he claims: "The drive to be an I is at the same time the drive to think and to feel" (Hardenberg 2003 [1795/1796], FS 32, 24). Compare the expression in *Fichte Study* 23: "The drive to be I—the first. I is I" (Hardenberg 2003 [1795/1796], FS 23, 19). In the *Fichte Studies*, the question of an interrelation between thinking and feeling takes on immense significance that goes far beyond the determination that Fichte and Kant gave it. In his moral philosophy, Kant is not interested in a detailed theory of feeling. Like the Stoics, he maintains that thinking and insight must find maxims of action that satisfy the requirements of the law of freedom and the categorical imperative. In the *Critique of the Power of Judgment*, Kant develops a threefold theory of feeling: (1) pure sensual and subjective private feelings, (2) moral feelings such as respect, generated by reason, and (3) feelings of beauty and the sublime, which are a kind of intelligible feelings, generated by reflecting power of judgement (Waibel 2017, 38–39). In Fichte's *Wissenschaftslehre*, one important function of feeling is to indicate the existence of the Not-I, that is, of the external world, in the context and as a consequence of Fichte's theory of drives (Fichte 1965, 2021a [1794/1795], GA I 2, 429; SW 1, 301). Another function of feeling is to announce approval or disapproval (Fichte 1965, 2021a [1794/1795], GA I 2, 448; SW 1, 325). Although these Fichtean functions of feeling play a role in Hardenberg's *Fichte Studies*, in this chapter I concentrate on another aspect of feeling, namely the aspect of indicating the becoming conscious of what was previously unconscious in the mind or the original action, the "Urhandlung".

The first time we encounter the idea of drive in the *Fichte Studies* is in a negative way. In *Fichte Study* 16 we find that "no special drive grounds intuition". We read further that intuition is situated between feeling and reflection and, it is said, to some extent surprisingly, is caused by them:

> The faculty of intuition. No special drive grounds intuition. Intuition is divided for feeling and reflection. Without application it is one. Applied, it

> is *tendency* and *product*. The tendency belongs to feeling, the product to reflection. The subjective belongs to feeling, the objective to reflection.
>
> /Relation between faculty [*Vermögen*] and power [*Kraft*]/
>
> Feeling and reflection together cause intuition. It is the unifying third thing—that however cannot enter into reflection and feeling—because substance can never creep into accident, synthesis can never quite appear in thesis and antithesis./Thus an object arises out of the interaction of two non-objects. Application to the original act. (Hardenberg 2003 [1795/1796], FS 16, 13)

The idea that intuition (*Anschauung*) is caused by feeling and reflection is a surprising alternative to the theories of Kant and of Fichte. It was Friedrich Schiller in his essay *On the Aesthetic Education of Men in a Series of Letters* (1795), who introduced a formal and a material drive that are bridged by the play drive (*Spieltrieb*). Schiller emphasised that the latter is not an independent drive like the two other drives. The fact that Hardenberg does not refer to intuition as a special drive, but as something between tendency—that is feeling—and product—that is reflection—reminds us of Schillers systematic structure. Moreover, Hardenberg asks for the "[a]pplication to the original act". We have to keep this in mind.

In the next *Fichte Study* 17, he speaks of the formal drive associated with reflection. Its equivalent, the material drive, is not explicitly mentioned, however, although the term *matter* often comes along with the term *form* in the *Fichte Studies:*

> If feeling is there in consciousness and it is supposed to be reflected, which [reflection] is caused by the drive to form [*Formtrieb*], then a mediating intuition must precede, which itself must be produced by a preceding feeling and a preceding reflection, which however cannot enter into consciousness—and the product of this intuition now becomes the object of the reflection. This *appears* now, however, to be progression from un-limited to limited [from reflection to feeling] and is *actually* precisely a reversed progression. (Hardenberg 2003 [1795/1796], FS 17, 14)

In the practical part of the *Foundation of the Entire Wissenschaftslehre* of 1794/1795, Fichte does acknowledge striving and drive as crucial elements of epistemology with regard to explicating the intentionality of thought, but he never explicitly mentions a formal drive. Hardenberg may have become acquainted with the notion of a formal and material drive by studying the ideas of two other prominent contemporary thinkers.

On the one hand, Karl Leonhard Reinhold starts his third book of his *Essay on a New Theory of the Human Capacity for Representation* as follows: "drive for representation in general", which is divided into the two basic drives, the "drive for material" and the "drive for form" (Reinhold 2013 [1789], OA 561/356). Note the reference to Reinhold regarding Hardenberg's considerations. In *Fichte Study* 19 he introduces the matter of feeling (though not as a material drive) and explicitly mentions "drive in general" and two other, not further defined "known drives", in *Fichte Study* 22. Here he elaborates: "The application of the original act to intellectual intuition occurs through drive in general. Because drive in general is the union of both known drives" (Hardenberg 2003 [1795/1796], FS 22, 18).

It should also be briefly mentioned that, on the other hand, Schiller also uses Reinhold's terminology of material drive and formal drive in his *On the Aesthetic Education of Men in a Series of Letters* (Schiller 1962, 1967 [1795], 13th and 14th Letter, *NA 20, 347–360 and 85*–99) and analyses Fichte to advance his thoughts for his own purposes. Hardenberg is particularly concerned with understanding and opening up the core of thinking, reflecting and philosophising with the concepts of drive, feeling, and original act, which are not identical but very closely connected

He uses the conception of drive to describe the living principle of the entry of the original act into the agency of the absolute I. Fichte's theorems are arranged in a remarkable way and are used to resolve the questions raised by Hardenberg. He writes in *Fichte Study* 23:

> At first, as we know, the original act constitutes itself. Then on the strength of its self-constitution it must turn to the "Something". But we have just seen the original act first constitute itself as the opposite of intellectual intuition, then progress to intellectual intuition—all on the strength of the drive of the original act. [...]
>
> Here, original act, in relation to the absolute I.
> There, only as half, here as whole—there relatively real—here absolutely [real]—there as form alone—here as form and matter at once—namely, through drive.

> /About drive? The drive to be I—the first. I is I. That is drive, and everything/
>
> That which is *something* for the *original act alone*, is subject and object for the original act in the I.
>
> /I *want* to call the original act alone a relative original act, and the original act in the I an absolute original act./
>
> Intellectual intuition is also something. (Hardenberg 2003 [1795/1796], FS 23, 19)

Hardenberg obviously understands here the drive of the original act and the original act itself as the element that explains to some extent the transition from being into consciousness, the transition from not-being-conscious of oneself into being-conscious of oneself. He describes this transition as a living principle that connects both.

Fichte uses the term drive in the *Foundation of the Entire Wissenschaftslehre* as a principle of self-conscious subjectivity striving for reason. For Fichte, the subject's striving and its concrete form of realisation as drive is an intellectual principle of life. This is one of the elements in Fichte's theory that does not originate from Kant's theory at all. It is part of the early theory of intentionality, which makes an explicit appearance in Fichte's "Foundation of the Science of the Practical" (Fichte 1965, 2021b [1794/1795], GA I 2, 385–451).

Spinoza incorporated a theory of drive into his system before Fichte did. It remains uncertain whether Hardenberg knew Spinoza's *Ethics* in depth. What is certain, however, is that Spinoza's theory of conatus, that is, drive, in the third part of the *Ethics* is a theory that describes the drive as the decisive principle of life (Spinoza [1677] 1999, 238–247, 336–371). A principle that belongs both to the world of extended bodies in time and space and to the world of spirit. It represents the same drive, namely that all appearances of the Spinozian cosmos are sometimes an appearance of the physical world and sometimes an appearance of the spiritual world (Waibel 2012; Damasio 2003, 2005 [englische und deutsche Version]; Kisser 2002). Whether or not Hardenberg was familiar with the dual nature of the Spinozian drive, it is clear that he uses it in a systematically comparable way. Drives are dynamics of life that belong to the physical world, but which can also be expressed as dynamics of freedom and subjectivity. It is this principle that provides at least a *quid facti* explanation of how it is possible to have an original act guiding from being into consciousness.

Hardenberg deals intensively with the development, constitution, and emergence of spiritual activities. A number of studies is summarised under the title "Genesis of the Drive" starting in *Fichte Study* 44, but already in *Fichte Study* 47 he no longer speaks about a duality—but a trinity of drives, powers and faculties. Hardenberg seems to have in mind that differentiation emerges in an originally indifferent unity. He thinks of drive's genesis as an action, a motion, a change, which is the equivalent to stasis, rest, and identity. It sounds as if he is giving us a definition of drive when he writes in *Fichte Study* 44: "A continuation in itself is a drive". Here he obviously means the continuation of motion. He attributes motion to the subject and identity to the (absolute) I. The *Fichte Study* goes as follows:

> Genesis of the Drive
> [44.] The I is absolutely One—the subject is divided absolutely—Interaction of the I in itself—It wants to be One, it wants to be divided. In the pure I alone both are absolute—the character of the absolute is: no change—no opposition—no *continuation*—stasis—rest—identity—Its character must be a manifold in relation to the subject—activity conditioned by rest.
> Yet the subject, the *ordine inverso* of the I, must have the character of the divided, the continuing, the active—in relation to the object its character must be that of rest conditioned by activity, oneness [conditioned by] manifoldness.
> /The character of pure act is this unconditioned alternation in the I/
> The activity of both divided things in the subject is separated, as long as the subject is not related [to the object], because since the subject is composed only of them [both divided things] they must both be active if the subject is active.—As soon as the one activity is related to another in an object—then it can be nothing other than a striving. A striving to be one—through intellectual power of vision—which unites subject and pure I in the object—Striving towards rest—but precisely for this reason an endless striving, so long as the subject does not become a pure I—which will not likely happen as long as the I is an I.

> This would be the drive, and indeed, the drive to be an I.
> A continuation in itself is a drive. Now since every activity is a continuation, there must be a drive in both members of the subject because they are active—these drives must thus be the drive to think and the drive to feel. (Hardenberg 2003 [1795/1796], FS 44, 31/32)

Based on Fichte, Hardenberg here ascribes indifferent identity as a starting point to the absolute I. Difference is attributed to the subject, as Hardenberg defines it here in his terminology. These different instantiations of the subject and their opposing predicates, here rest, standstill, there activity and motion, are the basis for explaining the emergency of drive's action. Hardenberg writes that there is an interrelation between the two of them: "Yet the subject, the *ordine inverso* of the I". The interrelation, that is, *ordo inversus*, also determines drives. When there is a drive, an action, then feeling and thinking also occur, as it will be shown in a moment. Just as with motion—the action of the subject—comes division and separation, so at the same time there is the striving for calm and identity. Hardenberg attributes this to the power of vision, or more precisely to the intellectual power of vision: "A striving to be one—through intellectual power of vision—which unites subject and pure I in the object—Striving towards rest" (Hardenberg 2003 [1795/1796], FS 44, 32). What Hardenberg develops here can be read as an explanation of the earlier idea in *Fichte Study* 16 where he claims that there is no drive of intuition.

In *Fichte Study* 47 there is a change—because he now writes that there are three drives, faculties and powers:

> I think—I feel—already with this the absolute subject's inner relationship—whole and part at once—is expressed. Both, reflection and feeling, are faculty and power, depending on whether the subject feels or thinks. If it feels, it is object, pure I—if it thinks, it is subject, divided I. The unity that accompanies it everywhere, that it is *there* completely, *where* it is—this is the highest, most essential character of its subjectivity.
> Thus the subject has
> three drives.
> Drive—to intuit. The vision-drive.
> Drive—to think. The thought-drive.
> Drive—to feel. The feeling-drive.
> Three faculties.

Faculty of vision.
Faculty of thought.
Faculty of feeling.
Three powers.
Power of vision.
Power of thought.
Power of feeling. (Hardenberg 2003 [1795/1796], FS 47, 34)

One cannot always consider Hardenberg's thoughts as systematically consistent. Here, in the first section, he identifies the pure I with feeling, the subject with thinking, and then goes on to claim the three different drives, faculties and powers of intuition, thinking and feeling, but does not develop this new scheme in detail.

Here and there he rather tries out thoughts and sometimes they turn out to be successful and sometimes a failure. It is up to the reader to judge. Nevertheless, we can state that the original act generally brings about an initial need to reflect, to philosophise. However, this need, as long as it is not reflected upon, is a kind of feeling that draws attention to a lack and is not necessarily and immediately understood as a feeling. If the interrelation between feeling and reflection occurs and with it the necessary separation of the two, the drive for identity is ignited. With this, the original act and the intellectual "power of vision", or more precisely, intellectual intuition, appear on a higher level.

Interrelation, the Fichtean notion of *Wechselbestimmung* is very dominant in the *Fichte Studies*, especially the interrelation of feeling and thinking, the feeling of the absolute or pure I and the thinking of the subject. Hardenberg coined another, now famous terminology, that of *ordo inversus* or *inverso ordine*. This figure of thoughts sometimes appears explicitly, sometimes implicitly. What this has to do with drives will be examined in the next section.

2 Interrelation/*Ordo Inversus* of Thinking and Feeling

When Hardenberg speaks of the "genesis of the drive" in *Fichte Study* 44, he is addressing the absolute I in its relation to the relative I that he also calls the subject. This reminds us once again of *Fichte Study* 32. Its title is

"On the empirical I" and the opening question is: "How does the absolute I become an empirical I?" (Hardenberg 2003 [1795/1796], FS 32, 24). It is clear that the absolute I in Hardenberg's reflections is an instance of isolation and abstraction but should not be taken in a literal sense as an "Absolute". It is the instance of a normative ideal and law, including the laws of thought that appears in the empirical subject. Hardenberg notes: "The absolute I is at once both a united and a divided thing" (Hardenberg [1795/1796] 2003, FS 32, 25).

At the end of these reflections, he speaks of a "famous conflict" that characterises the theory of the I:

> The pure I is only divided insofar as it is one, and is only one insofar as it is divided.
>
> This is, however, the famous conflict within the I—which constitutes its character—which is already to be found in the absolute original act—and which is nothing but a necessary deception of the mediated I alone—that wants to stop being mediated and thus strives against itself. The conflict is, as conflict, merely in the mediated I and is necessary precisely because it is not originally a conflict—one need only pay heed to the inverted order [*ordo inversus*] of the mediated I—because this is the actual ground of the contradiction [cf. #36]. (Hardenberg 2003 [1795/1796], FS 32, 25/26)

This is the first time in the *Fichte Studies* where Hardenberg speaks not only of a contradiction [*Widerstreit, Wechselbestimmung*] but of the *ordo inversus*. On the basis of his previously developed interrelation between feeling and reflection, which is located at a higher level of complexity compared to intellectual intuition and original act, Hardenberg feels justified in also assuming the relationship between the empirical I and the absolute I according to the very same structure, provided the absolute I is thought of as an independent original form that correlates with the relative form determined by matter.

In its relation to the empirical I, the absolute I is only mediated: "Thus there are two mediated I's at hand—the I that is felt and the I that is thought. The absolute I goes [according to feeling, VLW] from the infinite to the finite, the mediated I [according to reflection, VLW] from the finite to the infinite" (Hardenberg 2003 [1795/1796], FS 32, 25). So when the absolute I interacts with the empirical I, Hardenberg again has to ask himself the following question: "But how does the absolute I arrive at the finite, where it then is a mediated I, according to its own laws?" (Hardenberg 2003 [1795/1796], FS 32, 25).

The considerations Hardenberg comes up with after revisiting his question are outlined in the second part of the short text and make clear that he understands the absolute I as a mental unity, which represents the indeterminate ideal of pure reason and its legitimacy.

For the question relevant here, Hardenberg uses the context of the absolute I, which is similar to Fichte's thinking in the *Foundation of the Entire Wissenschaftslehre* and therefore writes: "the undetermined ideal of the pure I is thus characteristic of the I in general" (Hardenberg 2003 [1795/1796], FS 32, 25). Hardenberg mentions a second characteristic of the I: "The pure I is only divided insofar as it is one, and is only one insofar as it is divided. This is, however, the famous conflict within the I—which constitutes its character—" as an *ordo inversus* (Hardenberg 2003 [1795/1796], FS 32, 25/26).

In Hardenberg's *Fichte Studies*, the notion of feeling is immensely upgraded, as has already been shown, because it describes the original character of philosophy. Using the adjective "original" reminds us of philosophy's very beginning in ontogenetic and phylogenetic terms. It reminds us of the translation of the Greek word *Philosophy* meaning "the love of wisdom". It reminds us of enthusiasm, of life as a driving force, of the feeling of freedom to pursue truth and the striving and wanting to find it.

Even his frequently asked question of what philosophy is, is driven by the question and method of philosophy from the earlier non-philosophical understanding of the world. Although feeling and its varieties are often mentioned, one does not find the quality of feeling specified. Its most striking definition is being the antipode of reflection, thought and knowledge. Illuminating and contemplating this antithesis is the subject of an abundance of notes in the *Fichte Studies.*

In *Fichte Study* 15, Hardenberg asks: "What is then a feeling?" The answer, which is thought through within the *Fichte Studies* and also in numerous other passages, seeks to find a determination of the relative indeterminacy of feeling, although, as Hardenberg insists, this specific determination at the same time negates the essence of feeling. Hardenberg adds: "It can only be observed in reflection—the spirit of feeling is then gone" (Hardenberg 2003 [1795/1796], FS 15, 13).

Here Hardenberg makes one of his fundamental propositions that shows the importance he attaches to understanding the method associated with the *ordo inversus*. Moreover, in *Fichte Study* 15, he writes: "Feeling cannot feel itself". This is a self-evident statement. One has a feeling of whatever kind, but one does not feel the feeling. This sentence again shows that a feeling needs another of its kind in order to be grasped, understood, and even given a specification. It holds also true that the feeling loses something of its original quality as soon as one tries to keep it in mind, to transform its fleeting appearance and to think about it or examine it.

Hardenberg does use a feeling to describe the beginning of philosophising, but he does not define it. The feeling that points to the need, even the drive, to philosophise is therefore not given a specification. Astonishment, however, is a feeling that one is accustomed to associate with the origin and drive of philosophising. However, this is not what Hardenberg seems to be talking about here.

In *Fichte Study* 19, Hardenberg writes only: "Reflection finds the need of philosophy, or of an assumed systematic connection between thinking and feeling—because the need is in feeling" (Hardenberg 2003 [1795/1796], FS 19, 14). The "need of philosophy" is understood as a feeling and so stands in contrast to reflection. Here, one might note that need could also be identified as drive; however, Hardenberg does not use this specific term here that he uses so often elsewhere. The feeling of need is obviously distinguished from the feeling of amazement here. Probably both moments are only two sides of the same coin. Amazement expresses the subject's affection, which is evoked by wonder or appreciation for an object. The object may be sensually given or ideally thought. Amazement, then, leads to a closer examination of the specific object and deeper engagement with it. In contrast, need is a subject-oriented feeling. Subjects expressing a need have the feeling of lacking something and it may be undetermined how this lack can be satisfied.

In *Fichte Study* 19, Hardenberg unites the back and forth direction of feeling and reflection in two schemes. The schemes, indicating the sequence of numbers 1 to 4, are to be read from left to right on the one hand and from right to left on the other. He calls the schematisation the "basis of all philosophizing". It is worth quoting this passage from the *Fichte Study* 19 in full length.

> The human being thinks and feels—it limits both *freely*—it is determined matter.
>
> /This would be Fichte's [notion of] intellect. The absolute I is this determined matter before the original act occurs in it, before reflection is applied to it/
>
> Thus we have observed the most natural path in our deduction of philosophy—The need for a philosophy in consciousness—apparent progress from limited to unlimited—reflection upon that—apparent progress from unlimited to limited—results of this reflection—results of the feeling of this reflection—reflection upon these results according to those results—Discovered connection, or philosophy.
>
> 1.   2.
> Feeling ——————— Reflection
> 4.   3.
>
> Reflection   Feeling
> *Back* and *forth* direction.
> Sphere exhausted—connection there.
> This is [the] basis of all philosophizing.
> 1. /So/   2.
> An *apparent* given ————— Why?
> Therefore   4.   /Thus/   3.
> That which is presented in the answer—Something found in us
>
> The connection between "So" and "therefore" is *mediated through* and *in* the subject—the sphere—the absolute or relative absolute. (Hardenberg 2003 [1795/1796], FS 19, 16)

It might come as a surprise that in the first scheme, the terms "feeling" and "reflection" are first arranged from left to right and then vice versa, that is, chiastically, and that numbering starts from left to right and then continues clockwise. The set of keywords from the two schemes that follow the same order in the preceding paragraph, and the complementary designations of the next sequence of numbers clarify them as follows. Philosophy begins with a need that appears in consciousness as a feeling, as something that attracts the mind to engage with it, which then leads to the reflection on the need. The result of this reflection is the reflected consciousness of the needy feeling. This in turn leads to a reflected awareness of the reflection itself. Reflection then turns out to be the unity that first establishes a connection that can only come about through thinking, that is, what Hardenberg, referring to Kant, calls "assumed systematic connection" (Hardenberg 2003 [1795/1796], FS 19, 14).

The two schemes as the "basis of all philosophizing", presented in a condensed and linear form, contain the following aspects:

1. Feeling: An *apparent* given.—So
2. Reflection:—Why?
3. Feeling: Thus—Something found in us
4. Reflection: Therefore—That which is presented in the answer (Hardenberg 2003 [1795/1796], FS 19, 16)

In Hardenberg's opinion, feeling, because of its non-thetic form, refers above all to the infinite, the absolute, but only becomes conscious as deprivation in reflection. Hence, feeling is precisely in opposition to reflection, as it is not intentional and not thetic. That is why, in the *ordo inversus*, feeling can be an indicating state referring to the absolute, which apparently becomes conscious in the process of reflection and is then revised in a second reflection.

The thoughts on "direction" of feeling and reflection, of freedom and determination are conceived as a movement swinging back and forth between these two moments. Hardenberg understands the relationship, as with many things, according to the structure of the *ordo inversus* (Frank and Kurz 1977; Molnár 1970). The systematic connection between feeling and reflection, constructed according to the *ordo inversus*, can be understood as the germ and the model of all original productivity through drives occurring in consciousness.

Not only the *ordo inversus* of feeling and reflection is fundamental to Hardenberg's philosophical conception, but also the object relation, insofar as the object is understood to be partially given to the subject and partially found or (as it is the case in other contexts) to be invented. The appearance of feeling in consciousness is, according to this scheme, "[a]n *apparent* given—So". The feeling appears as suddenly as the object of intuition. It exists without requiring conscious spontaneity, which usually accompanies thought as soon as the will actively directs its power to contemplation. But the feeling is, as here stated, only an "*apparent* given". The reason obviously lies in the fact that in feeling the object is given differently compared to intuition. The feeling that originally arises is a subject state, whose content and object are at first completely undetermined. The undetermined state is a so-being, which dares reflection to ask for the reason, the "Why" of the so-being. Reflection examines and scrutinises the undetermined state of feeling more closely. In reflection, the state of feeling is replaced by the realisation of the reason for feeling, namely "Thus—Something found in us". Hardenberg assumes that reflection obtains an answer to the question of reason. However, he does not elaborate on how it finds this answer. Nevertheless, the feeling certainly indicates that the "So" is to be found in the subject and that the "Why", which is a feeling, can be "presented in an answer". In reflecting on the feeling, one also learns to understand oneself and one's function in the indispensable interaction with the feeling. This structure is also the basis for the following statement:

> Emergence expresses in its ambiguity the highest philosophical truth—the surrender of the I, in order to complete itself—thus it is clear how, through its emergence, it emerges. It finds *itself, outside* itself [Ein-Innenfindung]. This finding becomes a finding out—a finding in reality, in theoretically determined reality, which is the only one there is for the I. (Hardenberg 2003 [1795/1796], FS 98, 48)

In this paragraph, Hardenberg very densely sums up his philosophical, systematic core idea, in which finding oneself through reflection is thought of as the origin, genesis and finally also the completion of the subject as well as of philosophy. Although in the preceding context there was only talk of finding, here, however, emergence is discussed in its ambiguity. The ambiguity of emergence means, on the one hand, the origin of expressing oneself by means of feeling and reflection, even if feeling is only indirectly addressed. On the other hand, it is understood as an ongoing process of emerging, of coming into being, and is completed in terms of quality by achieving excellence in one's life and in terms of quantity by one's oeuvre at the end of life.

The emergence of something new, something that did not exist before, is, however, not an entirely isolated process. It is at the same time a finding, albeit it is a finding outside itself, which Hardenberg calls "a finding out—a finding in reality". In the interrelation between emerging as well as finding out, and finding in, the close relationship between necessarily given and freely produced is present without going into detail. Although the relationship between feeling and thinking is not explicitly mentioned here, these thoughts are nevertheless systematically based on the *ordo inversus* of feeling and reflection. Hardenberg calls the original freedom, which is only open and accessible to feeling "mere-being—or chaos" (Hardenberg 2003 [1795/1796], FS 3, 6; Janke 1979).

"Should there be a still higher sphere, it would be the sphere between being and not-being.—The oscillating between the two.—Something inexpressible, and here we have the *concept of life*" (Hardenberg 2003 [1795/1796], FS 3, 6). Hardenberg calls its "tangibleness" the "highest *presentation* of the incomprehensible" (Hardenberg 2003 [1795/1796], FS 12, 11). We often encounter that the *ordo inversus* turns into paradoxes in the *Fichte Studies*. Hardenberg seems to truly love that kind of wording. He wants to uncover "life" and knows that it is ultimately impossible, as he reflects in *Fichte Study* 3:

> Here philosophy is at a standstill and must remain so—because life consists precisely in this, that it cannot be grasped. Philosophy can aim only at *being*. Human beings feel the boundary that circumscribes everything for them, for themselves, *the first act*; they must believe it, as certainly as they know everything else. Consequently we are here not yet transcendent, but rather in the I and for the I. (Hardenberg 2003 [1795/1796], FS 3, 6)

Not grasping, for Hardenberg, does not necessarily mean not grasping anything at all, because one can still feel. Feelings and sensations can be captured and held on to, thinking makes it possible to grasp and analyse them. However, the original character of feeling is destroyed in the process. In the process of destroying, what can be grasped is captured and then pinned down in a word. That happens, or seems to happen, in the conceptless judgement of the subject, who bases their judgements on feeling.

To this end, Hardenberg asks himself towards the end of *Fichte Study* 649 and towards the end of the *Fichte Studies* in general:

> *Pure sensation*, whence comes that which is the most certain touchstone of truth, etc.? Our sensation excludes and includes—but does not determine—In its totality it is the unconscious intuition of the world of spirit—we sense an objective whole—and with every determination of the faculty of knowledge we sense a certain right or wrong, without being able, without special activity of the imagination, etc., to state it. This relationship gives much excellent material to the faculty of judgment. (Hardenberg 2003 [1795/1796], FS 649, 188)

What matters to Hardenberg time and again is the origin of the formation of consciousness, because he has realised that thinking draws from the sources that feed it, but he cannot grasp them and thus gives feeling and sensation their own space within epistemology. This all can be perceived as a dimension of his philosophical explorations that is as innovative as it is lively. Creativity is not only one of his major philosophical topics, but with his reflections, his bold trains of thought and his jumping from thought to thought, Hardenberg himself orchestrates creativity and gives insight into his black box. The alternation between feeling and thinking, which Hardenberg thought to be fundamental, challenges the fully established philosophical rationality that focuses mainly on cognition. Consciousness is intuition and thinking. Consciousness, however, is not limited to that. It is much more. Consciousness allows its cognitive abilities to form higher powers. A space of what Kant calls the comprehensible is created and formed. In contrast to Kant, Hardenberg not only wants to show the boundaries of the comprehensible, he also wants to know the incomprehensible. Thus, he discovers that the boundaries between these two are not only virtual, not only a line, but that the boundary actually occupies an immense space, namely the space of feeling. Feeling draws attention to itself, in its apparent "mere being—or chaos" (Hardenberg [1795/1796] 2003, FS 3, 6) determinable relations can be observed.

Although there will always be boundaries where the incomprehensible begins, thought always finds ways guided by feeling and sensation, and makes connections and sheds light on "the unconscious intuition of the world of spirit" (Hardenberg 2003 [1795/1796], FS 649, 188). Hardenberg's thoughts convey the plasticity of consciousness in an intriguing way with a tangible and at the same time elusive vividness.

In conclusion, I'd like to share some thoughts on Hardenberg's *Monologue* and the drive of language. The drive of language—another drive that could be called an original act (*Urhandlung*)—reveals at least a drive of communicability and a drive of sociability. In his *Monologue*, Hardenberg highlights the ironic aspects of communication. Even here we encounter an *ordo inversus*. We think we are communicating content, ideas, but ultimately, and sometimes it is true, we are speaking for the sake of commonality with others.

Sometimes the drive of the I that we constitute succeeds in dominating its world; sometimes the drive of feeling and thinking dominates in Hardenberg's thought experiments on the origin of things that transcend being to come into consciousness.

3 Friedrich von Hardenberg/Novalis *Monologue* (1798/1799)

The monologue has received a great deal of academic attention and is one of the most successful prose writings by Friedrich von Hardenberg/Novalis. The monologue is provocative, bold, and very modern. Unfortunately, however, the manuscript hasn't survived, nor is the title certain, nor can its date of composition be precisely determined; rather, one must rely on the textual form of the first printing in *Novalis Schriften*, Dritter Theil (*Novalis Writings*, Part Three), edited by Ludwig Tieck and Eduard von Bülow, Berlin, 1846. In this text, Hardenberg speaks of "*Sprachtrieb*", a drive of language, an urge to speak, that, in the context of the entire textual style, is reminiscent of the *ordo inversus* from the

Fichte Studies, without being explicitly mentioned as such here. In order to make the drive of language addressed in this text clear, the *Monologue*, in Joyce Crick's translation, is reproduced in its entire length:

> Speaking and writing is a crazy state of affairs really; true conversation is just a game with words. It is amazing, the absurd error people make of imagining [92//93] they are speaking for the sake of things; no one knows the essential thing about language, that it is concerned only with itself. That is why it is such a marvellous and fruitful mystery—for if someone merely speaks for the sake of speaking, he utters the most splendid, original truths. But if he wants to talk about something definite, the whims of language make him say the most ridiculous false stuff. Hence the hatred that so many serious people have for language. They notice its waywardness, but they do not notice that the babbling they scorn is the infinitely serious side of language. If it were only possible to make people understand that it is the same with language as it is with mathematical formulae—they constitute a world in itself—their play is self-sufficient, they express nothing but their own marvellous nature, and this is the very reason why they are so expressive, why they are the mirror to the strange play of relationships among things. Only their freedom makes them members of nature, only in their free movements does the world-soul express itself and make them a delicate measure and a ground-plan of things. And so it is with language—the man who has a fine feeling for its tempo, its fingering, its musical spirit, who can hear with his inward ear the fine effects of its inner nature and raises his voice or hand accordingly, he shall surely be a prophet; on the other hand, the man who knows how to write truths like this, but lacks a feeling and an ear for language, will find language making a game of him, and will become a mockery to men, as Cassandra was to the Trojans. And though I believe that with these words I have delineated the nature and office of poetry as clearly as I can, all the same I know that no one can understand it, and what I have said is quite foolish because I wanted to say it, and that is no way for poetry to come about. But what if I were compelled to speak? what if this urge to speak [*Sprachtrieb, drive of language*, VLW] were the mark of the inspiration of language, the working of language within me? and my will only wanted to do what I had to do? Could this in the end, without my knowing or believing, be poetry? Could it make a mystery comprehensible to language? If so, would I be a writer by vocation, for after all, a writer is only someone inspired by language? (Hardenberg, [1798/1799] 1984, 92–93)

The highly ironic tone, the playfulness in dealing with language and the self-contradictory character of the thoughts on language open up the possibility of seeing the short text less as thoughts on language in general (O'Brien 1995, 195–198), sometimes also read in the perspective of Heidegger's understanding of language (Di Cesare 1995, 149–168), and more as a poetological contribution on creative, inventive, writerly language, as suggested further. According to this text, no one can really use language, because it is functionalised and instrumentalised, in order to communicate things. The *Monologue*, however, informs us that this is not the "unique characteristic of language". Hardenberg knew Fichte's writing of the *Grundrisses des Eigenthümlichen der Wissenschaftslehre (Outline of What Is Distinctive of the Wissenschaftslehre with Regard to the Theoretical Power)* from 1795, which he published after *the Grundlage der gesamten Wissenschaftslehre (Foundation of the Entire Wissenschaftslehre)* of 1794/1795, in order to present in it not so much what is unique, that is, special (*Eigenthümlich*), but rather what is concrete and particular in the *Wissenschaftslehre*, namely concrete sensation, concrete intuition and finally space and time. Hardenberg speaks of the "unique" (*dem Eigenthümlichen*) a good deal in the *Fichte Studies* of 1795/1796. One of the most succinct passages reads in *Fichte Studies* 285:

> Transitus vom homogenen zum heterogenen, oder vom Gemeinschaftlichen zum Eigenthümlichen/Allgemeinen, Besondern/ist Eine Art von Freyheit—
>
> Vom Eigenthümlichen zum Eigenthümlichen, heterogenen zum heterogenen—die Andre Art von Freyheit. (Hardenberg [1795/1796] 1965, FS 285, 204)

> Transition from homogeneous to heterogeneous, or from [what is] common to [what is] unique [*Eigenthümlich*] / universal, particular/ is a kind of freedom—
>
> From unique characteristic [*Eigenthümlich*] to unique characteristic [*Eigenthümlich*], heterogeneous to heterogeneous—[is] the other kind of freedom. (Hardenberg [1795/1796] 2003, FS 285, 101)

Here, the unique is contrasted with the common and general in order to recognise a kind of freedom in it. Another kind of freedom lies in the transition from the "unique characteristic to unique characteristic", which is at the same time called a relationship of the heterogeneous to the heterogeneous. It is obviously a special freedom of the unfamiliar to the unfamiliar.

In fact, at the beginning of the *Monologue* there is talk of language in general, and of the fact that whoever wants to communicate things with language calls language itself into question, or even overturns it: "But if he wants to talk about something definite, the whims of language make him say the most ridiculous false stuff" (Hardenberg [1798/1799] 1984, 93). But at least it seems to be a unique characteristic, a special feature of language, to be able to cancel out its most practiced function, its function of communication itself, as is ironically suggested here. This self-cancellation, this self-negation, however, functions in such a way that this is at the same time communicable to the reader and worthy of being communicated.

Once again, Hardenberg's play with negations is reminiscent of a poetic variant of what Georg Wilhelm Friedrich Hegel will later develop with his philosophy of double negativity. Negation is per se dependent on language, logic, thinking, and concepts. And so, it is also an essential characteristic of ironic speech that what is said is set and pronounced as well as questioned or even negated and overturned. In the *Monologue*, an analogy is drawn between the failure of speech and Cassandra's warnings of impending disaster. Cassandra's words, her warnings were laughed at and thrown to the wind, only to end up—which the text does not explicitly say, but the myth enthusiast knows—having been right after all, without being heard and taken seriously.

To annul the function of speaking and communicating through language, to negate it through speaking—to use language to communicate that language is meaningless—is nevertheless a way of speaking that is capable of making visible various acts of language, of speaking. Language that cancels itself brings out language in its uniqueness and primordiality (in other words, performative contradiction). Saying, speaking and writing are negated in order to arrive at a new, playful and free way of dealing with language. If at the beginning of the text there is talk of an indeterminate "speaking and writing [*Sprechen und Schreiben*]" as a "crazy state of affairs [*närrischen Sache*]", a "game with words [*Wortspiel*]", a speaking of everyone, of "people [*Leuten*]" who do not understand the "mystery [*Geheimniß*]" of language, accused by an unspecified speaker, then towards the end of the text the identity of the speaker changes to an I who, with the last words of the text, identifies himself as a "writer by vocation [*berufener Schriftsteller*]", as "someone inspired by language [*Sprachbegeisterter*]". This "I" credits himself with having grasped something of the "nature and office of poetry [*Wesen und Amt der Poesie*]" in order to state and articulate its knowledge that its words were "foolish [*albern*]", that "no one can understand [*kein Mensch verstehen kann*]" what is being talked about, and that anyway "is no way for poetry to come about [*so keine Poesie zu Stande kommt*]" (Hardenberg [1798/1799] 1984, 92–93).

This someone inspired by language, driven to speak and write by the "urge to speak", sees language as a self-referential, self-sufficient affair that turns the drive of language, the urge to speak back on itself by negating every form of intentional language action through positing, by depriving it of its value, and by declaring the adequacy of language to be able to achieve its goals null and void. Even this text itself, that is, the *Monologue*, is also held up by language, in a self-referential twist, as it is called, and thereby fails to achieve its goal, which is apparently quite essentially to live in and with the language in such a way that one "has a fine feeling for its tempo, its fingering, its musical spirit [*ein feines Gefühl ihrer Applikatur, ihres Takts, ihres musikalischen Geistes hat*]", that one "can hear with his inward ear the fine effects of its inner nature [*in sich das zarte Wirken ihrer innern Natur vernimmt*)" (Hardenberg [1798/1799] 1984, 93; see also Naumann 1990). Intentions for awakening a finer sense of speech are expressed and at the same time extinguished in order to reach the origin of speech through negation. It becomes apparent to me that here, once again, the *ordo inversus* as a play of negativity is present, something that Hardenberg had already reflected on so eloquently in the *Fichte Studies*, so as to no longer merely elaborate the original action ("Urhandlung") of all feeling and thinking in its freedom, but to lead us to the original action of language itself. The "urge to speak [*Sprachtrieb zu sprechen*]", it is said, is "the mark of the inspiration of language, the working of language [*Kennzeichen der Eingebung der Sprache, der Wirksamkeit der Sprache*]". Those who are inspired by language use a technique of multiple negation and ironisation to reach the source of non-intentional, uncontrolled language creation, as it were—out of the nothingness of the unconscious, and in this way allow language itself to act, to have its say in a free play (Hardenberg [1798/1799] 1984, 93).

Translation: Linnea Gustavsson and Julia Herndlhofer.

References

Damasio, Antonio R. 2003. *Looking for Spinoza: Joy, Sorrow, and the Feeling Brain*. London: Heinemann.

———. 2005. *Der Spinoza-Effekt: Wie Gefühle unser Leben bestimmen*. Trans. Hainer Kober. Berlin: Ullstein.

Di Cesare, Donatella. 1995. Anmerkungen zu Novalis' Monolog. *Athenäum. Jahrbuch für Romantik* 5: 149–168. https://edoc.hu-berlin.de/bitstream/handle/18452/6293/di-cesare.pdf?sequence=1.

Eckhardt, Hans-Wilhelm. 1987. *"Wünsche und Begehrungen sind Flügel": Die Genesis der Utopie bei Novalis*. Frankfurt am Main, Bern, New York, Paris: Lang.

Fichte, Johann Gottlieb. 1965 [1794/1795]. Grundlage der gesammten Wissenschaftslehre als Handschrift für seine Zuhörer 1794/95. In *Gesamtausgabe der Bayerischen Akademie der Wissenschaften* (GA), ed. Erich Fuchs, Hans Gliwitzky, Reinhard Lauth, and Peter K. Schneider, vol. I 2: *Werke 1793–1795*, ed. Reinhard Lauth, and Hans Jacob, with the assistance of Manfred Zahn, 173–460. Stuttgart-Bad Cannstatt: frommann-holzboog.

———. 2005 [1798]. *The System of Ethics: According to the Principles of the Wissenschaftslehre*. Trans. and Ed. Daniel Breazeale, and Günter Zöller, Cambridge: Cambridge University Press.

———. 2021a. Foundation of the Entire Wissenschaftslehre." In *Foundation of the Entire Wissenschaftslehre and Related Writings (1794–95)*, Johann Gottlieb Fichte, Ed. and Trans. Daniel Breazeale, 195–380. Oxford: Oxford University Press.

———. 2021b. Outline of What Is Distinctive of the Wissenschaftslehre with Regard to the Theoretical Power. In *Foundation of the Entire Wissenschaftslehre and Related Writings (1794–95)*, Johann Gottlieb Fichte, Ed. and Trans. Daniel Breazeale, 195–380. Oxford: Oxford University Press.

Frank, Manfred, and Gerhard Kurz. 1977. Ordo Inversus: Zu einer Reflexionsfigur bei Novalis, Hölderlin, Kleist und Kafka. In *Geist und Zeichen: Festschrift für Arthur Henkel*, ed. Herbert Anton, Bernhard Gajek, and Peter Pfaff, 75–97. Heidelberg: Winter.

Freud, Sigmund. [10]2003 [1915]. Das Unbewußte (1915). In *Das Ich und das Es: Metapsychologische Schriften: Einleitung von Alex Holder*, ed. Sigmund Freud, 117–169. Frankfurt am Main: Fischer.

Hardenberg, Friedrich von (Novalis). 1965 [1795/1796]. Abteilung II: Philosophische Studien der Jahre 1795/96 (Fichte-Studien). In *Novalis*

Schriften: Die Werke Friedrich von Hardenbergs: historisch-kritische Ausgabe in 4 Bänden, einem Materialband und einem Ergänzungsband mit dem dichterischen Jugendnachlaß und weiteren neu aufgetauchten Handschriften, ed. zweite nach den Handschriften ergänzte, erweiterte und verbesserte Auflage, begründet von Paul Kluckhohn und Richard Samuel, herausgegeben von Richard Samuel in Zusammenarbeit mit Hans Joachim Mähl und Gerhard Schulz, vol. II, Einleitung von Hans Joachim Mähl 29–103, Text 104–296, Lesarten 693–724, Anmerkungen 724–727. Stuttgart, Berlin, Köln: Kohlhammer.

———. 1984 [1798]. "Monologue" [translation by Joyce Crick]. In *German Aesthetic and Literary Criticism. The Romantic Ironists and Goethe*, ed. Kathleen M. Wheeler, 92–93. Cambridge: Cambridge University Press.

———. 2003 [1795/1796]. *Fichte Studies.* Ed. Jane Kneller. Cambridge: Cambridge University Press.

Janke, Wolfgang. 1979. Enttönter Gesang—Sprache und Wahrheit in den Fichte-Studien des Novalis. In *Erneuerung der Transzendentalphilosophie im Anschluß an Kant und Fichte*, ed. Klaus Hammacher and Albert Mues, 168–203. Stuttgart-Bad Cannstatt: frommann-holzboog.

Kisser, Thomas. 2002. Affektenlehre als Ethik: Spinozas Begriff des conatus und die Konzeption menschlichen Handelns. In *Spinozas Lehre im Kontext*, ed. Achim Engstler and Robert Schnepf, 215–244. Hildesheim, Zürich, New York: Olms.

Mähl, Hans Joachim. 1965. Fichte-Studien: Einleitung. In *Novalis Schriften: Die Werke Friedrich von Hardenbergs: historisch-kritische Ausgabe in 4 Bänden, einem Materialband und einem Ergänzungsband mit dem dichterischen Jugendnachlaß und weiteren neu aufgetauchten Handschriften*, vol. II, 29–103. Stuttgart, Berlin, Köln: Kohlhammer.

Molnár, Géza van. 1970. *Novalis "Fichte Studies": The Foundation of his Aesthetics.* The Hague: Moutoun.

Naumann, Barbara. 1990. *Musikalisches Ideen-Instrument: Das Musikalische in Poetik und Sprachtheorie der Frühromantik*. Stuttgart: Metzler.

O'Brien, William Arctander. 1995. *Novalis, Signs of Revolution.* Durham: Duke University Press. https://hdl.handle.net/2027/mdp.49015002293141.

Reinhold, Karl Leonhard. 1789. *Versuch einer neuen Theorie des menschlichen Vorstellungsvermögens.* Prag, Jena: Widtmann und Mauke.

———. 2013 [1789]. *Gesammelte Schriften: Kommentierte Ausgabe, vol. 1: Versuch einer neuen Theorie des menschlichen Vorstellungsvermögens*, ed. Martin Bondeli, and Silvan Imhof, Basel: Schwabe.

Schiller, Friedrich. 1962 [1795]. Ueber die ästhetische Erziehung des Menschen in einer Reihe von Briefen (1795). In *Schillers Werke: Nationalausgabe, vol. 20: Philosophische Schriften. Erster Teil*, ed. Benno von Wiese, with the assistance of Helmut Koopmann, 308–412. Weimar: Verlag Hermann Böhlaus Nachfolger Weimar.

———. 1967. *On the Aesthetic Education of Man in a Series of Letters*. Trans. Elizabeth Mary Wilkinson, and Leonard Ashley Willoughby. Oxford: Oxford University Press.

Spinoza, Baruch de. 1999 [1677]. *Ethica Ordine Geometrico demonstrata et in et in quinque Partes distincta. Ethik in geometrischer Ordnung dargestellt und gegliedert in fünf Teile.* New translation, edited and with an introduction by Wolfgang Bartuschat. Latin—German. Hamburg: Felix Meiner.

Waibel, Violetta L. 2005. Unbewusstes oder Unaufrichtigkeit? Freud und Sartre im Widerstreit über Reflexivität und Präreflexivität des Bewusstseins. In *Anatomie der Subjektivität: Bewusstsein, Selbstbewusstsein und Selbstgefühl*, ed. Thomas Grundmann, Frank Hofmann, Catrin Misselhorn, Violetta L. Waibel, and Véronique Zanetti, 460–491. Frankfurt am Main: Suhrkamp.

———. 2012. Philosophieren als Weg: Anmerkungen zu Spinoza und Fichte mit einem Exkurs zu Hölderlin. In *Affektenlehre und amor Dei intellectualis: Die Rezeption Spinozas im Deutschen Idealismus, in der Romantik und in der Gegenwart*, ed. Violetta L. Waibel, with the assistance of Max Brinnich, and Peter Gaitsch, 200–230. Hamburg: Meiner.

———. 2017. Metaphysik des Schönen und Erhabenen im Hinblick auf das Tragische bei Kant, Schiller und Hölderlin. In *Kant und die Folgen: Die Herausforderung in Ästhetik, Ethik, Religionsphilosophie: Wiener Jahrbuch der Philosophie XLVIII/2016*, ed. Rudolf Langthaler and Michael Hofer, 34–101. Wien: New Academic Press.

12

Drive, Will, and Reason: Reinhold and Schiller on Realizing Freedom after Kant

Jörg Noller

1 Introduction

From a philosophical point of view, the concept of drive (*Trieb*) is far from being clear. The conceptual field of drive includes other concepts such as power or force (*Kraft*), incentive (*Antrieb*), inclination (*Neigung*), propensity (*Hang*), motive (*Motiv*), striving (*Streben*), spontaneity (*Spontaneität*), and will (*Wille*). As such, the concept of drive is a constitutive moment for a philosophical theory of human freedom, since it concerns the transition from mere thoughts to real actions. In other words, the concept of drive serves the purpose to *realize* and *express* individual human freedom. However, there is an ambivalence entailed in the concept of drive. For we speak of "drive" in two senses that seem to contradict each other. On the one hand, we use the word "drive" to refer to states in which we are overwhelmed ("driven") by motives that are not in our control. On the other hand, the word "drive" or "driving force"

J. Noller (✉)
University of Munich, Munich, Germany
e-mail: joerg.noller@lrz.uni-muenchen.de

M. Kisner, J. Noller (eds.), *The Concept of Drive in Classical German Philosophy*,
https://doi.org/10.1007/978-3-030-84160-7_12

(*Triebfeder*) can be used to denote our *own motive* to realize a certain action, and to express our individuality.

This ambivalence between autonomy and heteronomy that is contained in the concept of drive can be found in Kant's philosophy. For on the one hand, a drive is, according to Kant, something that belongs to our empirical nature and is therefore heteronomous. A driving force, on the contrary, can be something autonomous, since according to Kant's ethics it is pure reason that becomes practical in the form of the incentive of the moral feeling of respect. Karl Leonhard Reinhold's and Friedrich Schiller's conception of drive can be interpreted as a systematic response to the ambivalence in Kant's use of the concepts of drive (*Trieb*) and driving force (*Triebfeder*). Both Reinhold and Schiller attempt to integrate the notion of drive into a comprehensive theory of freedom, and to interpret the concept of drive not in terms of heteronomy but in terms of autonomy as volitional self-determination. Both attempt to develop a unified account of human action of which drives are a constitutive element. As such, they develop a compatibilist account that reconciles nature and freedom, sensibility and reason. Thereby, both thinkers interpret Kant's ethics in terms of their theory of drives and transform it at the same time.

My chapter is structured as follows: I will first outline the ambivalence in Kant's use of the concept of drive between autonomy and heteronomy of the will. I will then show how Reinhold and Schiller attempted to develop a unified account of drive that overcomes the Kantian distinction between autonomy and heteronomy. Whereas Reinhold conceives of the drives in terms of unity, Schiller interprets them in terms of harmony. Finally, I analyze their transformation of Kant's ethics, focusing on the relation between drive(s) and the moral law.

2 Kant on Drive and Driving Force

Kant uses the word "drive" in different contexts in his philosophical work.[1] We can distinguish between two main contexts, namely (1) autonomy and (2) heteronomy. The notion of drive often comes along in the context of the concept of "Triebfeder", which can be translated in terms

[1] For a discussion of the concept of drive in Kant's work, see Heidemann and Kisner in this volume.

of "incentive", "spring", "motive" or, more literally, "driving force".[2] Kant defines "Triebfeder" as "the subjective determining ground of the will of a being whose reason does not by its nature necessarily conform with the objective law" (Kant 1997, 5:72).[3] He thereby distinguishes the notion of will from the notion of driving force or incentive:

> [T]he will stands between its a priori principle, which is formal, and its a posteriori incentive (*Triebfeder*), which is material, as at a crossroads; and since it must still be determined by something, it must be determined by the formal principle of volition as such when an action is done from duty, where every material principle has been withdrawn from it. (Kant 1998a, 4:400)

According to this double-aspect theory, and following the terminology of Frankfurt (1971), the human will can be determined either empirically, by the law of nature, that is, on the basis of first-order desires, or by the moral law, that is by pure practical reason, or self-reflected second-order volitions. In each case, it is the same will, although distinguished in the *lower* and the *higher* faculty of desire (Kant 1997, 5:25). The former concerns what Kant calls "animal choice" (*arbitrium brutum*), the latter our "free choice" (*arbitrium liberum*) insofar as it is determined by pure practical reason alone (Kant 1998b, 6:213). Accordingly, Kant distinguishes between a negative concept of freedom as the "independence from being *determined* by sensible impulses" and a positive concept as "the capacity of pure reason to be of itself practical" (Kant 1998b, 6:214). This entails the "reciprocity thesis" (Allison 1986), according to which "a free will and a will under moral laws are one and the same" (Kant 1998a, 4:447).

Kant argues that the human will needs to be motivated in order to lead to an action, and he therefore considers the problem of rational moral motivation as one of the great questions in practical philosophy (see Noller 2019b). In the Introduction to his *Critique of Practical Reason*, he raises the question of "whether pure reason of itself alone suffices to

[2] The Cambridge Edition of Kant's works translates "Triebfeder" with "incentive", thereby following Mary Gregor's translation.

[3] Kant's works are cited by the number of the volume of the *Akademie-Ausgabe*, which contains the cited text, a colon, and the number of the page(s) in this order in the whole text. The (German) *Akademie-Ausgabe* is listed as Kant (1900, ff). in the bibliography at the end. The translations cited from are listed as Kant (1997, 1998a, b).

determine the will or whether it can be a determining ground of the will only as empirically conditioned" (Kant 1997, 5:16). Of special importance for his theory of rational moral motivation is Kant's conception of the moral feeling of respect or reverence (*Achtung*). The moral feeling of respect for the moral law is "the sole and also the undoubted moral incentive (Triebfeder)" (Kant 1997, 5:78), as Kant puts it.

However, Kant's conception of respect as the only moral incentive leads to what Henry Allison has called "Reinhold's Dilemma" (Allison 1986, 422; see also Noller 2019a, 855). For only in being motivated by the incentive of respect we seem to be autonomous, whereas in acting evilly we seem to be bound to our sensible nature. As a reaction, Karl Leonhard Reinhold in his *Letters on the Kantian Philosophy* highlights some problematic implications that seem to follow from Kant's theory of the incentive of respect. The "friends of Kantian Philosophy", as Reinhold puts it, "attempted to save the will from the slavery of the instinct only in that way insofar as they made it the slave of the force of reason". These "friends", according to Reinhold, "attempted to escape will's necessitation of sensibility only by conceiving the will as being inevitably necessitated by reason" (Reinhold 2008, 200). Following to this conception, however, "a moral action could only be understood as a mere effect of reason" (Reinhold 2008, 200). "If the will were only free with regard to moral actions, and the ground of *immoral* actions laid outside the will in external obstacles and barriers", Reinhold goes on, "so also the reason of moral actions would by no means to be found in the mere self-activity of practical reason but rather in the absence of these obstacles that are entirely independent from this reason" (Reinhold 2008, 200). So the question arises of how we can act autonomously both on the basis of empirical *and* rational incentives.

3 Reinhold and the Unity of Drives

Reinhold develops his conception of drive especially in his 7th *Letter on the Kantian Philosophy*, which deals with the "previously unrecognized difference between the unselfish (*uneigennütziger*) and the selfish (*eigennütziger*) drive, and between these two drives and the will" (Reinhold

2008, 161). He argues that this distinction is "[a] difference from which, once it has been admitted, all the basic doctrines and subsequent clauses in that account [...] result" (Reinhold 2008, 161). What are these basic doctrines in Reinhold's account? By means of his conception of drive, Reinhold reconceives (1) Kant's notion of will, (2) his account of morality, (3) his conception of moral motivation, and (4) his theory of freedom as autonomy (see Noller 2019a, 2020a).

In contrast to the tradition following Wolff and Baumgarten, Reinhold analyzes the will not in an upper and a lower faculty of desire, but "in two *original*, essentially different and essentially united drives". According to Reinhold, the selfish drive is "founded in *sensuality*, and has *pleasure* as an object *at all*". The unselfish drive, in opposition, is "present in personal *self-activity*" and "establishes a law that is only necessary by itself" (Reinhold 2008, 134). Reinhold calls these two drives "driving forces" (*Triebfedern*) of the will, "in how far they are engaged in the arbitrary satisfactions or dissatisfactions of the desire (*Begehren*)" (Reinhold 2008, 178). Using the terminology of Frankfurt (1971), we can interpret a person's desire (*Begehren*) in terms of first-order desires, and both the selfish *and* the unselfish drive as second-order volitions: "The demand of the selfish as well as the unselfish drive are prescriptions which are given to the person, one by pure practical reason, the other by theoretical reason, mediated by pleasure and displeasure" (Reinhold 2008, 179). Both drives can be understood as different types of the use of reason—as theoretical reason, which, as instrumental reason, satisfies the selfish drive by applying the relation of end and means to objects, thus, viewing the world solely from one's *own perspective*, and as practical, which satisfies the unselfish drive by opening up a normative dimension and by taking the interests and perspectives of other freely acting persons as the basis for individual decision. The unselfish drive can thus be understood as a historically and socially situated and materialized form of the moral law, which is thereby transformed into a real structure. At the same time, however, the demand of the selfish drive is not to be identified with the heteronomy of a lower faculty of desire, but is an equally possible option of free self-determination. On the basis of his reinterpretation of Kant's notion of drive and driving force, Reinhold reformulates Kant's categorical imperative, and puts it as follows: "*In all your actions of will, the satisfaction or dissatisfaction of your selfish drive be subordinated to the demand of the unselfish one*" (Reinhold 2008, 139).

By reconceiving the notion of *Trieb* and *Triebfeder*, Reinhold argues for a compatibilism of freedom and nature. The selfish and the unselfish drive both are possible motives of freedom, since they "must of course be present in every willing (*Wollen*), and are, because without them no will is conceivable, *reasons*, and because the *objects* of will are determined by them, *objective* reasons of the will" (Reinhold 2008, 181). Reinhold argues that both moral and immoral action is an expression of freedom:

> In moral action, absolute practical necessity and freedom are united in so far as the absolutely necessary law, i.e. the effect of practical reason, is executed by choice in a given case, and so far made the effect of freedom. In immoral action, natural necessity and freedom are in so far united as the requirement of the selfish drive, which is only in accordance with the laws of nature of desire, but contradicts practical law, is carried out by choice, and is in so far elevated to the effect of freedom. (Reinhold 2008, 201)

According to Reinhold, the will, or more precisely, the faculty of choice (*Willkür*), concerns the level of reflexive attitudes toward the first-order desires, insofar as the person can "determine herself to satisfy or not satisfy a desire" (Reinhold 2008, 174). Both drives are thus considered options of freedom. They constitute the basis of freedom, which Reinhold calls "morality" in a broad sense: "Morality in a *broad sense*, that is, in an act of will", concerns "the relationship between the demands of the selfish and unselfish drive" (Reinhold 2008, 138). The decision for freedom arises from the specific combination of both drives in an integrated will: In moral action, as Reinhold puts it,

> *absolute practical* necessity and *freedom* are *united* insofar [emphasis J. N.] as the absolutely necessary law, the effect of practical reason, is carried out by choice (*Willkür*) in a given case, and is made so far away the effect of freedom. In immoral action *natural* necessity and *freedom* are *united* insofar [emphasis J. N.] as the demand of the selfish drive, which is merely in accordance with the natural law of desire, but contradicts practical law, is carried out by choice (*Willkür*). (Reinhold 2008, 201)

The Kantian idea of an autonomy of reason is thus reconceived by Reinhold, for it is no longer pure practical reason alone that is capable of directly determining the will. Or, as Reinhold puts it: "The reality of the satisfaction of the selfish drive no longer depends on this [unselfish] drive alone" (Reinhold 2008, 197 f.). Therefore, positive freedom consists, according to Reinhold, "in the self-activity of the person […], a very special self-activity, which must be distinguished from the self-activity of reason" (Reinhold 2008, 192). Freedom, as Reinhold reconceives it, consists in the relationship of the person to her own nature, which is actively included in the individual decision of the will: "Both drives are essential to human nature, and the unselfish [drive] can by no means do without the selfish in order to satisfy its own demands, to fulfill the moral law" (Reinhold 2008, 141). In view of this connection of both drives in the individual person, freedom of the will consists in the reflective arrangement of their will, so that free will can be understood as a personally *integrated* network of drives. Both the selfish and the unselfish drive are related to one another in the act of freedom and brought to a decision by being brought into a unity: "Willing, then, is not a mere expression of neither the selfish nor the unselfish drive […]; but it is self-determination for or against the demand of the unselfish drive, for the satisfaction or non-satisfaction of a demand of the selfish one" (Reinhold 2008, 135 f.). Through this indissoluble interweaving of both drives in the act of freedom, Reinhold's difference to Kant's concept of pure will becomes obvious. Freedom of will is possible only on the basis of the whole nature of man, which is expressed by the unity of both drives: "Since the will is the very ability to determine itself for the satisfaction or non-satisfaction of a demand of the selfish drive; and since these demands depend directly or indirectly on experience: so *all will* is empirical in this regard" (Reinhold 2008, 188).

By referring to the concept of drive, Reinhold's philosophy can be understood as an attempt to *realize* Kant's theory of freedom, which is not to be confused with mere moral psychology. Rather, Reinhold is concerned with an analysis of the will and the *entire* use of individual freedom.

4 Schiller and the Harmony of Drives

Already in his early dissertation *On the Connection between the Animal Nature of Man and his Spiritual Nature* (1780), Schiller developed a model of the unity of human drives, which he conceived as a harmony of man's mixed nature:

> The perfection of man lies in the exercise of his powers through contemplation of the plan of the world; and since there must be the most exact harmony between the measure of the power and the purpose on which it acts, perfection will consist in the highest possible activity of his powers and their mutual subordination. (Schiller 1962a; NA 20:50)[4]

In his *Aesthetic Letters*, Schiller argues against Kant's separation between pure rational form and sensible matter. For according to Kant, "[O]ne easily falls into thinking of material things as nothing but an obstacle, and of imagining that our sensuous nature, just because it happens to be a hindrance in this operation, must of necessity be in conflict with reason" (Schiller 1967, 87 n; NA 20:348 n.). By means of his concept of beauty, Schiller attempts to reconcile human nature in terms of freedom of the will (Noller 2020a, 2020b). It is in the case of beauty that a human person becomes perfectly free and that this freedom corresponds to her mixed nature and demonstrates her unity. Whereas Kant developed his conception of freedom as autonomy from the "fact of reason" (Kant 1997, 5:31), Schiller's theory, as it were, draws on the "fact of beauty". This conception of beauty serves as the formal condition for specifically personal freedom of the will. Schiller argues that only in a certain harmonious relationship between nature and reason can a reflexive volitional activity arise, since an imbalance would not permit the self-distancing of the will with regard to its first-order desires: "Exclusive domination by either of his two basic drives is for him a state of constraint and violence,

[4] The *Nationalausgabe* of Schiller's works is cited as NA followed by the number of the volume, a colon, and the number of the page(s) in this order in the whole text. If I refer to no other edition of the text, I translated the text myself. The texts in the *Nationalausgabe*, which are referred to, are listed as Schiller (1780) 1962a, Schiller (1790–1794) 1992, Schiller (1793) 1962b and Schiller (1795) 1967 in the bibliography at the end.

and freedom lies only in the co-operation of both his natures" (Schiller 1967, 119; NA 20:365). In order to be free, human persons demand "an intimate agreement between their two natures, of always being a harmonious whole" (Schiller 2005, 154; NA 20:289). This "intimate agreement" can be understood in terms of harmonious structure of first-order desires and second-order volitions in which the individual "resonance frequency" of the person is realized.

The human will is divided into two drives that "exhaust our concept of humanity" (Schiller 1967, 185; NA 20:347). Schiller's distinction of these volitional structures does not follow the Kantian opposition between autonomy and heteronomy. Rather, Schiller conceives of both drives as the basis for the realization of individual freedom.

> I have no qualms about using this expression [scil. "drive"] collectively, both for that which seeks to follow a law and for that which seeks to satisfy a need, although it is otherwise restricted to the latter only. Just as rational ideas become imperatives or duties as soon as they are placed within the limits of time, so these duties become impulses as soon as they are related to something specific and real. [...] This drive [scil. the form-drive] necessarily arises, and is also not absent from the one who acts against it. Without it there would be no morally evil, and consequently no morally good will. (Schiller 1962c–1963; NA 21:243–244, my translation).

How can we understand both drives of human nature? Schiller describes the complex personal drive structure as follows: The "sensuous drive" (Schiller 1967, 139; NA 20:374), or the "material drive"—as Schiller reformulates the finite and empirical nature of the human person—"proceeds from the physical existence of man, or his sensuous nature" and "set[s] him within the limits of time" (Schiller 1967, 79; NA 20:344). Here the material drive, as it were, "presses for reality of existence" (Schiller 1967, 81; NA 20:345). Each drive is characterized by its specific intentionality and its object. Schiller calls the object of the material drive "life, in the widest sense of this term" (Schiller 1967, 101; NA 20:355). As a "life impulse" (*Lebenstrieb*) (Schiller 1967, 139; NA 20:374), this drive represents the volitional structure directed toward the preservation of the individual. While the material drive constitutes the empirical and finite dimension of the person, the formal drive has "form" as its object, inasmuch as it "includes all the formal qualities of things and all the relations of these to our thinking faculties" (Schiller 1967, 101; NA 20:355).

However, both drives are not only in a synchronous relationship of coordination, but also in a genetic one of evolution. Schiller conceives a developmental theory of human freedom on the basis of nature: "The sensuous drive awakens with our experience of life (with the beginning of our individuality); the rational drive, with our experience of law (with the beginning of our personality); and only at this point, when both have come into existence, is the basis of man's humanity established" (Schiller 1967, 137; NA 20:373). Schiller speaks of a "priority of the sensuous drive" and argues that it "provides the clue to the whole history of human freedom" (Schiller 1967, 141; NA 20:374). Freedom, as Schiller puts it, "arises only when man is a complete being, when both his fundamental drives are fully developed" (Schiller 1967, 139; NA 20:374). Nature thus represents the real basis of freedom, out of which, as the last stage of volitional evolution, human freedom emerges. Schiller therefore neither harshly opposes nature to the individual person (here the problem of intelligible fatalism looms), nor does he allow it to merge completely into nature (here the problem of natural determinism and indifferentism occurs).

Since the form-drive and the material drive are directed toward "opposite ends", they "cancel each other out, and the will maintains perfect freedom between them" (Schiller 1967, 135; NA 20:371). This freedom, however, is just the freedom of indifference. Schiller understands the mutual relation of both drives in an individual decision of the person not as a mere opposition or equilibrium, but as a harmonious coordination, which he calls "play-drive". In the dynamic state of play-drive, the form-drive and the material drive "act in concert": "[I]t will, therefore, since it annuls all contingency, annul all constraint too, and set man free both physically and morally" (Schiller 1967, 97; NA 20:354).

The state of such a harmoniously integrated will, in which first-order desires harmonize with second-order volitions, is "to be looked upon as a State of Supreme Reality (*höchste Realität*), once we have due regard to the absence of all limitation and to the sum total of the powers which are conjointly active within it" (Schiller 1967, 151; NA 20:379), so that "through the use of his freedom"—through specific spontaneity—"it is now up to the mind (*Geist*) to make use of its tools" (Schiller 2005, 133; NA 20:263). The play or the game—both is expressed by the German word "Spiel"—is not so much a "limitation" but rather an "expansion" of the human person (Schiller 1967, 195; NA 20:358), or as Schiller puts it in his famous dictum: "[M]an only plays when he is in the fullest sense of the word a human being, and he is only fully a human being when he plays" (Schiller 1967, 107; NA 20:359).

Playing, however, does not mean a meaningless or indifferent activity. Rather, in playing the rules of play serve as reasons for actions; they do not restrict but rather enable freedom. In playing, "the material constraint of natural laws and the spiritual constraint of moral laws" are abolished in an esthetic state of "higher concept of Necessity, which embraced both worlds at once; and it was only out of the perfect union of those two necessities that for them true Freedom could proceed from which" (Schiller 1967, 109; NA 20:359).

Schiller understands inner necessity not as a kind of fatalism that makes freedom impossible. Rather this kind of necessity is a state of "real and active determinability" (Schiller 1967, 141; NA 20:375), that is a unity of determination and contingency, which—in contrast to metaphysical or logical necessity—is an expression of individual freedom. In playing, the form-drive and the material drive are preserved, so that this state is "not just lawlessness but rather harmony of laws, not arbitrariness but supreme inner necessity" (Schiller 1967, 125; NA 20:367). Schiller distinguishes this free state of inner necessity from that of the necessitation of reason in Kant's conception of autonomy and respect by referring to Kant's notion of heautonomy. Schiller thus argues, as it were, with Kant against Kant, for the concept of heautonomy refers to the capacity of esthetic judgment: "The perfect can have autonomy insofar as its form is purely determined by its concept; but heautonomy is possible only in beauty, since only its form is determined by its inner essence" (Schiller 1992; 2003, 169; NA 26:210).

How can we better understand Schiller's concept of the play-drive in terms of freedom? In the equilibristic state of "real and active determinability" (Schiller 1967, 141; NA 20:375), the human will is not necessitated, but rather united, integrated and coordinated. The play-drive is a "middle disposition", "in which sense and reason are both active", and "in which the psyche is subject neither to physical nor to moral constraint, and yet is active in both these ways" (Schiller 1967, 141; NA 20:375). Therefore, according to Schiller, "contemplation (reflection)" is also "[t]he first liberal relation which man establishes with the universe around him"—or, as Schiller formulates the difference between first-order desires and second-order volitions: "If desire seizes directly upon its object, contemplation removes its object to a distance, and makes it into a true and inalienable possession" (Schiller 1967, 183; NA 20:394).

5 Conclusion

Immediately after Kant, the concept of drive undergoes a significant transformation. Both Reinhold and Schiller use this concept in order to realize Kant's moral conception of freedom and autonomy. This realization proceeds in two ways. First, they attempt to interpret immoral actions as manifestations of free choice and not of heteronomy, as Kant's reciprocity thesis seems to suggest. Second, they include the empirical and individual nature of man in terms of the selfish drive (Reinhold) and the material drive (Schiller) into their conception of freedom. Whereas Reinhold conceives of both drives as a unity insofar as both are likewise forms of freedom, Schiller interprets them in terms of a dialectical harmony by mediating and sublating them in the concept of the play-drive.

References

Allison, Henry E. 1986. Morality and Freedom: Kant's Reciprocity Thesis. *The Philosophical Review* 95 (3): 393–425. https://doi.org/10.2307/2185466.

Frankfurt, Harry. 1971. Freedom of the Will and the Concept of a Person. *The Journal of Philosophy* 68 (1): 5–20. https://doi.org/10.2307/2024717.

Kant, Immanuel. 1900. *ff. Gesammelte Schriften,* edited by the Royal Prussian (later German) Academy of Sciences. Berlin: De Gruyter.

———. 1997. *Critique of Practical Reason.* Trans. Mary Gregor. Cambridge: Cambridge University Press.

———. 1998a. *Groundwork of the Metaphysics of Morals.* Trans. Mary Gregor. Cambridge: Cambridge University Press.

———. 1998b. *Religion within the Boundaries of Mere Reason.* Trans. Allen Wood and George di Giovanni. Cambridge: Cambridge University Press.

Noller, Jörg. 2019a. 'Practical Reason is not the Will': Kant and Reinhold's Dilemma. *European Journal of Philosophy* 27 (4): 852–864. https://doi.org/10.1111/ejop.12448.

———. 2019b. Reason's Feeling: A Systematic Reconstruction of Kant's Theory of Moral Respect. *SATS. Northern European Journal of Philosophy* 20 (1): 1–18. https://doi.org/10.1515/sats-2019-0012.

———. 2020a. From Autonomy to Heautonomy: Reinhold and Schiller on Practical Self-Determination. *Idealistic Studies* 50 (3): 261–274. https://doi.org/10.5840/idstudies2020821116.

———. 2020b. Heautonomy: Schiller on Freedom of the Will. *European Journal of Philosophy.* https://doi.org/10.1111/ejop.12576.

Reinhold, Karl Leonhard. 2008. *Briefe über die Kantische Philosophie: Zweiter Band.* Ed. Martin Bondeli. Basel: Schwabe (My translation).

Schiller, Friedrich. 1962a. Versuch über den Zusammenhang der thierischen Natur des Menschen mit seiner geistigen. In *Schillers Werke: Nationalausgabe, vol. 20, Philosophische Schriften: Erster Teil, Friedrich Schiller*, ed. Benno von Wiese, 37–75. Böhlau (My translation): Weimar.

———. 1962b. Ueber Anmuth und Würde. In *Schillers Werke: Nationalausgabe, vol. 20, Philosophische Schriften: Erster Teil, Friedrich Schiller*, ed. Benno von Wiese, 251–308. Böhlau: Weimar.

———. 1962c–1963. Ueber die ästhetische Erziehung des Menschen in einer Reihe von Briefen. In *Schillers Werke: Nationalausgabe*, vol. 20–21, *Philosophische Schriften*, Friedrich Schiller, edited by Benno von Wiese, vol. 20, 309–412, and vol. 21, 232–277. Weimar: Böhlau.

———. 1967. *On the Aesthetic Education of Man in a Series of Letters.* Ed. and Trans. Elizabeth M. Wilkinson and L.A. Willoughby. Oxford: Clarendon Press.

———. 1992. *Schillers Werke: Nationalausgabe*, vol. 26, *Briefwechsel: Schillers Briefe: 1.3.1790–17.5.1794.* Ed. Edith Nahler and Horst Nahler. Weimar: Böhlau.

———. 2003. Kallias or Concerning Beauty: Letters to Gottfried Körner. In *Classic and Romantic German Aesthetics*, ed. J.M. Bernstein, 145–183. Cambridge: Cambridge University Press.

———. 2005. "On Grace and Dignity." Translated by Jane V. Curran. In *Schiller's "On Grace and Dignity"*. In *Its Cultural Context: Essays and a New Translation*, ed. Jane V. Curran and Christophe Fricker, 123–170. Rochester (New York), Woodbridge (Suffolk): Boydell & Brewer.

13

Drives in Schelling: Drives as Cognitive Faculties

Paul Ziche

1 Preliminary Remarks: Drives Between the Animal and the Prototypical Human

"Triebe", "drives",[1] carry, to today's ears at least, the connotation of the reductively natural, of the lower, animal levels in humans, of that what is

This chapter builds forth upon results of my research project on "Structuring the world of ideas around 1800" that was financed by NWO, the "Dutch research council". Many thanks go to NWO and to my colleagues in this project.

[1] Etymologically, these terms are related (see the Grimms' dictionary (Grimm & Grimm 1854–1961) entry "drive"), so "drive" can be used as a fully adequate translation for "Trieb" here. Unless indicated otherwise, all translations in this chapter are my own. A nice illustration of the richness of the semantics of "drive" in this period can be found in Pénisson (2002).

P. Ziche (✉)
Department for Philosophy and Religious Studies, Utrecht University, Utrecht, Netherlands
e-mail: p.g.ziche@uu.nl

M. Kisner, J. Noller (eds.), *The Concept of Drive in Classical German Philosophy*,
https://doi.org/10.1007/978-3-030-84160-7_13

beyond our rational control.[2] In the period around 1800, we find a strong discursive line that positions "drives" in a radically different fashion. Terms like "Kunsttriebe", that is, drives that make animals perform feats that one would normally explain via the human capacity to produce works of technology or of art, or "Bildungstrieb", that is, a formative drive that is omnipresent in organic nature, go beyond these strictly layered hierarchies.[3] *Friedrich Wilhelm Joseph Schelling* takes this discussion an important step further: He uses the term "drives" prominently in contexts that deal with the highest *cognitive* achievements, in particular with philosophy itself and with science (in the broad sense of German "Wissenschaft").

This establishes a link between Schelling's investigations into the subrational dimensions in humans, which he also puts in "drive"-related terms such as "hunger" or "addiction"/"Sucht",[4] on the one hand, and the realm of cognitive practices on the other. This, in turn, will have implications for our understanding of the hierarchy of human faculties—essentially, this hierarchy will have to be either subverted, or radically flattened out. The hierarchy of human faculties is itself related to explanatory hierarchies: Typically, one will assume that explanations proceed in a hierarchical way, either deductively, from general principles to particular instantiations of these principles, or inductively, from the particular to the general. Schelling, therefore, needs to offer us a strategy, drawing upon the notion of "drives", that can arrive at an alternative understanding of human faculties and of the structure of explanation.

Schelling's intervention can make use of the rich, complex, and unsettled semantics of the term "Trieb"/"drive" in the period around 1800.

---

[2] On Schelling and the unconscious, see McGrath (2010), Fftyche (2012), on Schelling and Jung, see Barentsen (2015).

[3] This discourse itself is part of, and it engenders, a debate about whether we can ascribe to animals capacities that are analogous to human capacities, or that can be put in a continuous connection with higher capacities (e.g. AA I,9,1, 196–209; AA I,8, 31). Paradigm examples, repeated stereotypically in the debate, are spiderwebs, beehives, and beavers' lodges. On the context, see Richards (2002), as well as the articles by Zammito and Cooper in this volume.

[4] Frequently, the terms "hunger" and "addiction" are used in close connection. Some significant passages for these terms and for their connection: AA I,11,1, 66, 71; AA I,17, 133, 157, 169; SW I,7, 466–467; SW I,8, 235, 254, 266, 352–353, 355, 361, 378–379; SW II,1, 294, 462; SW II,2, 631; SW II,3, 206, 447; SW II,4, 271, 283. See also Berg (2003) on a comparison between Schelling and Schopenhauer that highlights the importance of drive-related terms.

Hermann Samuel Reimarus states explicitly, in the "Vorbericht" of his agenda-setting book on animal drives in general and on "Kunsttriebe" in particular, that the unclear ("unbestimmt und schwebend") semantics of "drive" is a deplorable state of affairs that he tries to repair (Reimarus 1790, IV–V). The semantic openness of "drive" manifests itself along three dimensions. Firstly, the term "drive" itself is embedded into, and coordinated with, a large number of related terms that in many cases are offered, without detailed semantic argument, as alternatives to this term; "drive" does not have a precisely determined meaning in this time. Secondly, a large number of specific sub-forms of "drives" are introduced in this period; and, thirdly and most interestingly, "drives" are frequently given the task of opening up a more *general* and more *open* perspective upon the human faculties: "Drives" go beyond established sub-divisions of these faculties, and thus introduces novel ways of organizing the system of the faculties at large.

1. The first point, the fact that the notion "drive" is semantically unsettled in this period, can easily be illustrated. Johann Friedrich Blumenbach's text on the "Bildungstrieb" in organic nature equates "drive" with a number of other terms, without feeling the need to be semantically precise when he talks about a "*drive*, (or tendency, or *endeavour*, however one wants to call it)" (Blumenbach 1781, 12). Kant's *Anthropology*, when discussing the "passions" in § 80, uses drive-terms, "instinct", "propensity"/"Hang", and "inclination"/"Neigung" (Kant [1800], 265, 325; Kant 2007, 367, 420), with subtle differences, but in close connection.
2. Equally, the emergence of a large number of ever more specific drives is directly evident. Kant provides examples, for instance, with the terms "sexual instinct"/"Begattungstrieb" and "parental instinct"/"Älterntrieb" in § 80 of his *Anthropology* (Kant [1800], 265; Kant 2007, 367); another example: the term "Kristallisationstrieb", for example, in Schelling (AA I,7, 340).[5]

[5] The abbreviation "AA" refers to the historical-critical edition of Schelling's works, "SW" to the edition in the *Sämmtliche Werke*. On the "Kristallisationstrieb", see also Lind (1992, 272) on the usage of this term and of drive-terms in general as non-causal descriptors in Jakob Friedrich Fries' writings.

3. Hermann Samuel Reimarus opens his book on animal drives with a very general definition of what "drive" means: Drives are "all natural efforts towards certain activities" ("alles natürliche Bemühen zu gewissen Handlungen", Reimarus 1790, V). "Certain" in this definition seems to indicate that drives are directed towards specific activities (such as a spider is always building a web and not some other structure). This specificity of drives, their certainty, can also be read as an epistemic statement: Drives are not vague, but clearly determined. When drives are transported into the human realm, this clear determination of their products is given up. Around 1800, human activities, including drive-based activities, are typically described as not being clearly determined, and as, in their openness, having integrative functions. Schiller's aesthetic terms of "Formtrieb" and "Spieltrieb" (Schiller [1801], letters 12 and 14; see also Noller, in this volume; Stiening 2002) and the term "Kunsttrieb" itself are prominent examples for this strategy. Some compound "drive"-terms explicitly take up the idea that drives can have an integrative function; Schelling's "Ergänzungstrieb" (on which more in section III) is an example, as is Schleiermacher's "Ausdehnungstrieb".[6] A term such as "Betrieb", referring in a general way to a comprehensive whole of activities, which is used in the opening passages of Reimarus' book, indicates a similarly integrative or generalized function of the term, thereby turning an everyday term into a rather more technical usage (Reimarus 1790, IV). Despite the naturalist connotations of the term "drive", therefore, drives can be used in this period to arrive at ever higher levels of abstraction that range far beyond what is traditionally viewed as natural or animal in humans. Interestingly, Reimarus ascribes drives in a cognitive function also to animals, when he introduces "Vorstellungstriebe", drives linked to (mental) representations. In the case of animals, this applies to their mode of sensory perception and

[6] In the secondary literature (cf. König 2016), this term is used to summarize Schleiermacher's more extensive description of the human soul as being a "product of two opposing drives", one of which is the "striving to attract everything that surrounds the soul towards itself", the other being the "longing ["Sehnsucht"] to extend its own inner self, starting from the inside, ever further" (Schleiermacher [1799], 191).

to the (indistinct) way in which animals mix up representations of the past and of the present (Reimarus 1790, V).[7]

"Drives" are prominently discussed in a number of philosophical contexts around 1800. Fichte defines "drive" as "[a] striving ['Streben'] that produces itself, that is ascertained, determined, something certain" (Fichte [1794/1795], 418; on Fichte's notion of "drive", see Jacobs 1967, Cesa 2002; on Fichte and Schelling, see Barnouw 1972), and gives this term an important role in the practical part of his *Wissenschaftslehre*. This passage echoes Reimarus' account of drives as being directed towards a specific result. What is important in Fichte's account is the emphasis on a drive's being autonomously produced: Drives do not function as direct reactions upon stimuli; drives produce reactions that are specific, that are adequate, but are not directly determined by a stimulus. In acting upon drives, the actor stands open to an infinity of stimuli, thus allowing for flexible, but everywhere reasonable reactions that are not deductively (or in immediate stimulus-response chains) determined by an input. This is a pattern that features prominently also in the philosophy of nature of this period. Two key features of the organic realm, "irritability" and "sensibility", both signify modes of reacting upon stimuli in a way that is not determined by the stimulus itself.[8] Drives typically manifest themselves, throughout these texts, in the form of opposing tendencies; see the fundamental drives in Schiller or in Schleiermacher, the two contrasting tendencies in Fichte, the prominence of equilibrium models in Kantian and Schellingian philosophy of nature. This feature again requires us to move beyond an understanding of drives in term of unidirectional reaction chains.

Schelling fully exploits these features of the semantics of "drives". Some concepts from the semantic field of "drives"—such as "Bildungstrieb"—occur prominently in Schelling's texts, and precisely where we would expect them, in the case of "Bildungstrieb": in his

[7] See also Baader (1809, 120): "Every drive already carries with itself its wisdom, understanding, or, as the ancients say, its productive wit ["Witz", cf. Kant [1800], § 54] (every pleasure its ruse), and thus is *Kunsttrieb*".

[8] On "irritability" and "sensibility", see, for example, Durner et al. (1994, 375–498).

philosophy of nature.[9] However, we can also reconstruct a significant link between "drives" and higher cognitive achievements in Schelling. The pertinent discussions are spread out throughout his oeuvre (which is in itself remarkable, given the controversial question of the coherence of Schelling's philosophy). This chapter will collect the relevant evidence, and take the link between "drives" and the practice of science as a strategy to this end.

2 Schelling on "Instinct": Instincts as Cognitive Faculties

Let's start by looking at Schelling's usage of the term "instinct", a term that is a more recent addition to German language than "drive", but which is strongly related to "drives".[10] Schelling uses the term "instinct" precisely in the way that has already been adumbrated for the meaning of "drive", that is, for describing successful, but immediate and involuntary, and thus also not deductively goal-directed activities—including the activities of giving argument and of acquiring scientific knowledge. In his *System of Transcendental Idealism*, he discusses the epistemologically problematic field of physiognomy via "a kind of universal instinct" that makes us view a person's external physiognomic features as a "visible expression" of inner qualities, or at the very least as a reason to assume, as a "Vermuthungsgrund", as "affording the presumption" that these features indicate a person's "talent and [...] character itself" (AA I,9,1, 250; Schelling 1978, 171). The term "Vermuthungsgrund" introduces an important epistemic qualification: We cannot arrive at any unequivocal conclusions about the connection between physiognomic features and inner qualities, but we can do more than just guess blindly. And: The ability to come to physiognomic assumptions is not a rare gift, it is

[9] Schelling's discussion of the specific terms "Bildungstrieb" and "Kunsttrieb" will not be analysed here; however, the cognitive programme that Schelling develops in using the term "drive" is perfectly compatible with the way how Schelling deals with these more specific issues.

[10] The Grimms' dictionary, interestingly, already has an entry "Instinkt", thereby showing that this term is no longer standardly perceived as stemming from foreign languages (Grimm & Grimm 1854–1961).

"universal"; it is generally distributed throughout humankind. In his Munich lectures *On the History of Modern Philosophy* from the 1830s,[11] Schelling makes a similar point, but now no longer in the epistemically precarious field of physiognomy—he now discusses Kant's entire philosophy in terms of an instinct that led Kant to his profound innovations in philosophy.[12] More specifically, he uses this terminology with specific reference to the way how Kant's philosophy was received and transformed into innovative forms of philosophizing.

Schelling states here that Kant gets a lot right, which is not a surprising point to make, and that Kant does so via an instinct——and that most definitely is a surprising claim. What Kant gets right, or what his successors constructively and successfully did with his ideas, came about without him having intended it (SW I,10, 73); a philosopher cannot determine the reception of his own ideas, just as he cannot clearly predict the course that his own ideas take within his own train of reasoning. What reads like a staple point in hermeneutics is proposed by Schelling as a fully "general observation" concerning "all human deeds", including the creation of great philosophy: the "real *importance*, i.e. [the] true effects are generally others than those which were intended or which relate ['stand in proportionate relation to' would be a better translation] to the means by which they are produced" (SW I,10, 73; Schelling 1994, 94). More specifically, Schelling uses the plant and its "Bildungstrieb" as a comparison for how philosophy, and Kant's philosophy in particular, proceeds: Philosophy does not come about via explicit knowledge as to how to proceed, but philosophy, nevertheless, remains an ordered and structured activity that remains recognizably philosophical everywhere. This feature of philosophy (and of human activity in general) can only be labelled as a form of "feeling", more precisely, as a form of feeling that, nevertheless, stands up to the epistemic standards of "certainty", or as a "drive":

> [There is] [o]ne system which forces its way through all the successive appearances to the last point of its transfiguration; as the plant which

[11] A. Bowie's translation gives 1833–1834 as the most probable date for the version in SW that is standardly used for this text (Schelling 1994).

[12] On the importance of inherently heuristic notions in Kant's oeuvre, see Ziche (2016).

> begins to grow does not know what point it will arrive at but it still has a sure [in German, Schelling uses "gewiß", which has precisely this epistemic connotation, and echoes the passage in Reimarus that has been quoted earlier] feeling of that point, and this feeling is what drives it, is what we call the *drive* in it. Thus even if in this whole succession nobody had a clear concept of the goal, everyone felt that it was necessary to reach some final point, and precisely this feeling, this drive, which came into philosophy with Kant, distinguished this epoch from all earlier ones. (SW I,10, 74n; Schelling 1994, 94–95)

Schelling had written about Kant's achievements in the same terms already earlier in his oeuvre, and then explicitly in terms of a "drive" that drove Kant towards totality: "Kant's spirit displayed, after all, a *natural and unstoppable drive towards totality*" (SW I,6, 7).[13] Taken together, Schelling makes some rather extraordinary claims in these passages. In pursuing innovation in philosophy and science, we (and that includes Kant himself as the key originator of this step) need to and indeed can go beyond the limitations that Kant imposes upon human reason as being dialectically incapable of grasping totalities. What in Kant remains a regulative idea now becomes incorporated into scientific practice, in the form of a "sure" feeling, of an instinct.[14] This optimism can make use of a phrase that Kant had used himself when he acknowledged a "drive towards science", a "Trieb zur Wissenschaft" in his *Anthropology* (Kant [1800], 325). This drive, too, is inherently universal. What Kant wants to highlight here is that no individual person can exhaust science, that science can only be pursued via humankind as a historically connected and continuously developing totality.

This way of talking about drives is remarkable, not only because it is Kant in particular who is described here as an instinct philosopher, but

---

[13] Schelling uses the same terminology in remarks upon the emergence of early Greek philosophy, and upon the links between philosophy and the mysteries: "From Pythagoras (and even farther back) down to Plato, philosophers felt [German original: "erkennt sich die Philosophie selbst als eine exotische Pflanze im griechischen Boden, ein Gefühl"] that they were an exotic growth on Greek soil. The feeling is expressed, for example, in the fact that all those who were initiated into higher doctrines, either by the wisdom of the early philosophers or by the mysteries, instinctively turned [German original: "ein Gefühl, das schon in dem allgemeinen Trieb sich ausdrückte"] to the East, the motherland of the Ideas" (SW I,5, 346; Schelling 1966, 145).

[14] But see Baader (1797, 36) on drives as regulative principles.

also because Schelling uses "instinct" here in a distinctively cognitive sense.[15] Instincts are seen as being conducive to epistemologically successful results. The key features of instincts (and drives) that allow for this are their universality (they are shared by all competent participants in the epistemic discourse) and their being undetermined. When incorporated into science and philosophy, they become the key ingredient of creativity and innovation.[16]

The movement towards totality is built in into the very definition of "instinct". More precisely, Schelling sees instincts as combining an integrative function with their being an origin of activity—being undetermined, not yet having a clear direction, is a condition for being a first origin. In a significant passage in his treatise on the *World-Soul*, Schelling describes "instincts" (and the "all-invigorating soul"), in a phrase that resonates with Kant's transcendental deduction, as the capacity that brings together a naturally provided manifold into a complete, and moreover: into an organically complete, whole: "So, in the natural being, the manifold collects itself finally in the *instinct*, i.e. in the all-invigorating soul, without whose incitation ['Antrieb', in a slight semantic variation of 'drive'] no totality would ever come about that is complete/perfect in itself" (AA I,6, 250). This quote employs a phrase that refers to a form of passive synthesis ("sammelt sich"), but this passive collection is in itself

[15] Because instincts can be "certain" and philosophically productive, Schelling can more frequently couple the term "drive" with positive epistemic characteristics constructively: "richtig fühlender Instinkt", an instinct that is, interestingly, "feeling correctly" (SW II,2, 105), or more simply a "correct instinct", a "richtiger Instinkt" (SW II,2, 355).

[16] Schelling goes to some lengths to provide a general definition of "drive" that clearly relates to this term's epistemic functionality. In his philosophy of nature, he keeps rather close to the biological context of this term: "Trieb zur Bewegung, durch Sensibilität bestimmt, ist *Instinct*" (AA I,6, 249), an "*instinct* is a drive towards motion that is determined by sensibility", here explicitly subordinated to "drive". In a later text, he gives a far more abstract definition of drives: "Bloß partiales Einsseyn des Subjekts und des objektiven Grundes (der Identität und der Totalität) ist Instinkt"/"instinct is a merely partial identity of the subject and the objective ground (of identity and totality)"; "Was im Instinkt des Thiers handelt, ist noch ein ganz Objektives, aber es ist als dieses Objektive, ohne den Charakter des Objektiven abzulegen, zugleich ein Subjektives, was auch nachher an den Thieren mit einem Schein wirklicher, ihnen eigner Vernunft täuscht", that is: "what is active in the instinct of an animal still is completely objective; but in its being objective it also is subjective, without thereby losing the character of being objective. This subjective aspect then becomes deceptive when animals appear to command a genuine form of reason that is peculiar to them" (SW I,6, 457, 459). When Schelling links up the notion of "drives" to human scientific practice, he explicitly goes beyond the epistemic limitations that these passages state.

active ("alles belebende [...] Seele") and then functions as the "drive" or "incitation" ("Antrieb") towards an organically perfected whole. It is this synthesizing function of the instinct that makes it possible for Schelling to call the instinct the animal's faculty of perception (AA I,6, 250).[17] Instincts thus have an integrative function that can support both actions and perceptions. With this in mind, it becomes less surprising that Schelling can rather frequently talk about instincts as being "general" (e.g. AA I,9,1, 250; a very similar passage SW I,8, 283), referring to assumptions that we get correct purely on the basis of instinct. This explains why the "instinct" can also feature prominently in analysing the production of art: Works of art are produced without consciousness, from instinct, without thereby becoming in any way deficient (AA I,9,1, 320; see also the important passage AA I,9,1, 156–158, where Schelling links up the notions of "drive" and "feeling"). The specific faculties that Schelling relates to the instincts and drives are, not surprisingly, *future-directed faculties*. The future can never be predicted with apodictic certainty, as in the passages on physiognomy referred to earlier; still, we can relate to the future with some degree of anticipation, via a "presentiment that is of the kind of an instinct" (AA I,11,1, 115, "instinctartiges Vorgefühl"). This form of presentiment does not give full certainty, and still needs to be actualized.[18] In his *Stuttgarter Privatvorlesungen*, he uses related terms (e.g. AA II,8, 130: "praesagia"), and these faculty terms become related to another important dispositional term, that of character (as has already been the case with the remarks about physiognomy above). These terms are explicitly charged with the task of bringing the two realms of the spiritual and the natural, the—traditionally—higher and lower, more closely together.

[17] Interestingly, Schelling, in this context, also gives a qualified discussion as to whether, and to which extent, animals are able to have even "sublime representations"—not quite, but there also is no categorical difference in the way their cognitive apparatus operates as compared to higher cognitive capacities.

[18] Schelling uses a beautifully metaphorical term here. The "presentiment", which in this passage is anticipatively related to a threatening destruction, will "ausbrechen" if it is faced with the right kind of stimulus. "Ausbrechen" can be rendered as "escape from", but also as "starting to blossom", "break out into full bloom", mark the beginning of a new season or epoch, "erupt" with the violent force of a volcano.

Schelling uses drives prominently to describe a state in which we experience a disposition towards action, towards future developments, without experiencing an explicit goal. The fact that this goal is not available as an explicitly determined final state is captured by viewing this goal as a form of totality, in accordance with, and at the same time transcending, the Kantian limitations of the scope of human reason. When Kant discusses the possibility and the epistemic status of a systematic understanding of nature under empirical laws in his third *Critique*, he consistently uses faculty terms that are related to the future-directed terms that we also encounter in Schelling. Drives reach out into the future, which, of course, remains epistemically undetermined—as a consequence, drives are conceived of as inherently creative ways of determining a future state.

3 The Cognitive Function of Drives: "General" Drives and "Drives" Towards Science

In Schelling's usage of the terms, drives, just as instincts, can acquire higher, cognitive functions, and they can be general or universal in a number of respects: General in the sense of the objectivity-guaranteeing role of transcendental faculties in Kant; general in the sense of having the power to synthesize a manifold of input data; general in the sense of going beyond reductively egoist or subject-centred attitudes. Schiller's "Formtrieb" is a clear illustration of the higher functions that drives can acquire. The "Formtrieb" is directed against "feeling" and a merely sensuous "inclination"; where the "Formtrieb" takes a leading function, "[our] being is extended to the utmost extent", "da ist die höchste Erweiterung des Seins" (Schiller [1801], 472). Schleiermacher's drive towards extending our own self in a non-egoist, communicative way similarly illustrates the subject-transcending character of these higher drives. In terms such as "Gemeinsinn" or "Gemeingefühl" (see also Heidemann in this volume) we find similar extensions of a subjective faculty into a universally shared feature of humankind. As in the model of two opposing forces in Fichte or in his own philosophy of nature, Schelling discusses both "selbstische", ego-directed, centripetal drives and drives that are directed towards totality, "social drives" (SW I,5, 316). Humans are characterized by this overcoming of self-centredness. This is interesting in itself: It reads as an anti-idealist move, more specifically an anti-Fichtean move. But it is important to note that Schelling does not mean that the self-centredness disappears entirely; what he wants to see is that a focus on the ego "simultaneously" becomes "beautiful" (a comparison with Hegel's notion of "aufheben" might be interesting here), as a later passage makes clear:

> In humans, however, with the centre entering in an absolute fashion into the human being, the centrifugal tendency has reappropriated ["wieder in sich aufgenommen"] the centripetal tendency; what in animals was nothing but egoistic ["selbstisch"], now, in its being egoistic, simultaneously becomes beautiful and acquires a value on its own. (SW I,6, 488; see also SW I,6, 187 on the "Trieb zur Selbstheit"; AA II,6,1, 301–302; AA I,12,1, 166)

Schelling coins a term that translates the tendency towards totality into a rather technical term, and, consistently with everything that has been said so far, this has to be a form of drive, more precisely an "Ergänzungstrieb", a drive towards completion, towards incorporating the complementary parts into a totality (AA I,12,1, 166; SW I,6, 323; on this term, see, e.g., also von Hagen 1808, 129). This particular drive is not adequately understood as a drive towards just finding missing pieces that complete an as yet incomplete original state; its aim is the totality itself in which the, in itself deficient, origin of the drive and the complementary completion that it drives at come together. The key paradigms for this drive can be found in a cosmology that is inspired by Plato's *Timaeus*, and—again with Platonic connotations—in the erotic and sexual drives.[19]

Drives transcend ego-centred subjectivity; they aim at comprehensive totality; they are inherently creative. In a number of significant passages,

[19] Cf. also Oken (1843, 387), where the "Ergänzungstrieb" is identified with the "Geschlechtstrieb". In Schelling, cf. AA I,7, 103–104, where he describes the "Geschlechtstrieb" as a predecessor to the "Kunsttrieb". This ordering is as much an ordering in which higher levels replace the lower or earlier ones, as one where the higher is coming forth from, and incorporating, the lower.

Schelling uses these features of instincts and drives to characterize the proper attitude towards the practice of science. Kant's "drive towards science" has a parallel in Schelling's phrase of a "Trieb zur Erforschung", a drive towards research (AA I,12,1, 472). This is directly plausible: Research is aimed at knowledge that stands up to the highest standards of scrutiny, but research cannot arrive there by already knowing what the result of the activity of research can or will be (not even what it should be!); research does not proceed via deduction, it needs to stand open for surprise and creativity. According to the opening sentences of Schelling's 1803 text on the methodology of *University studies*,

> It is at the very beginning of his university career that the young man first comes into contact with the world of science. The more taste and inclination ["Sinn und Trieb"] he has for totality [the translation has "science" here, but this does not capture the German original, "das Ganze"], the more likely it is that this world will strike him as a chaos, a confused mass, a vast ocean upon which he is launched without star or compass. (SW I,5, 211; Schelling 1966, 5)

Schelling here sketches a rather bleak picture of the situation of the beginning student—this situation only becomes the bleaker the higher developed the student's "sense" or "drive towards totality" is. The picture that he suggests here, and that is confirmed by other significant passages that use the same terminology, is a two-layered one: "sense" and "drive" stand at the beginning of academic endeavours, but they lead the young practitioner into the openness of an ocean without direction. This experience of chaos needs to be brought under control, but in the beginning, it is important to keep the world of possible investigation as open as possible. "Sense" and "drive" set the chaotic driving forces in motion without which science would be impossible.

This may sound like, again, ascribing to the drives a lower, quasi-animal function. What is important to note here, however, is that chaos, that what traditionally has been thought to be lower, animal, has to be brought under control, but that this does not mean that it needs to be trimmed away; to the opposite: The chaotic state is a necessary precondition for scientific research. For Schelling, it is the function of philosophy itself to stimulate the drive towards totality that leaves the neophyte hapless in the face of chaos (this can explain why he captures Kant's innovations in instinct- and drive-terms)—philosophy is necessarily required in order to get the sciences going. Philosophy here gets the function of initiating a process to an unpredictable goal, or rather: philosophy has to make sure that the process of doing research, of acquiring knowledge, departs in and from a situation of full openness. What Schelling claims here is not the guiding function of a regulative idea that directs research towards a goal, even if this goal is characterized as the cognitively inaccessible goal of totality: The necessary openness is already built in into the starting point for research. This he makes perfectly plain in a number of passages in his lectures on *University Studies* (that these ideas have been important to Schelling is made evident by an almost literally identical echo of this passage in his *Aphorisms as an Introduction into the Philosophy of Nature*):

> Philosophy, by its nature, strives for the encompassing, the universal. In the individual or in a whole race, once the universal spirit of higher science and enlightenment by the Ideas is associated with the most living and diverse cognition of the particular, then we have the condition for cultural harmony such as gives rise [a beautiful biological metaphor in the German original: "erwachsen"] to healthy, straightforward [German original: "gerade", which is not adequately translated by the connotations of "being plain" that come with "straightforward"], productive knowledge and action. However, when striving for the encompassing and the universal, such as philosophy stimulates, is balanced neither by the richness of classical culture nor by genuine experience based on the observation of nature, the lopsided structure must sooner or later topple. Philosophy is not to blame for such a misfortune, but rather the shortcomings of its counterpart, which together with philosophy represents the complete cultural organism. (SW I,5, 262n; Schelling 1966, 56–57. The parallel passage in SW I,7, 137)

Philosophy incites the drive "towards the encompassing and the general", and this is a consistently positive and constructive role that philosophy has to play—the failures in cognitive endeavours cannot be ascribed to philosophy. What Schelling describes here closely resembles what Kant says about the interaction between *genius* and taste in the *Critique of the Power of Judgment*: Taste needs to keep unruly genius under control in order to arrive at beautiful works of art. In the passages just quoted, Schelling puts philosophy in the place of artistic genius, *and* he thereby also gives science the status of being a product of genius, as is art. This is a remarkable step, given Kant's stern insistence on the impossibility of giving genius a role in science—in these considerations, for the first time, a conceptual link is forged between science and the work of genius (Kant 1790, 308–309; Kant 2000, 187). Schelling is one of the key figures in this conceptual process that, by now, has become so deeply engrained in our thinking about genius that we tend to overlook how recent the link between "genius" and science actually is.[20]

A programmatic passage from the very opening of his *Aphorisms as an introduction into the Philosophy of Nature* takes up the same idea. Schelling here sticks to the terminology of a "drive towards generality", and to the problem of how this drive can be balanced and controlled so as to lead to stable results. Again, it is science that drives towards totality; and what he invokes to keep science in check here is religion—in their interplay, they can create a unity (a unity that Schelling describes with the strong term of a "heiliges Band", the sacred link, with its strong Platonic connotations) that can be described in terms of art and the beautiful (SW I,7, 141).

Schelling takes this yet a step further. He establishes as strong as possible a parallel between the inner structure of reality on the one hand, and the scientific attitude in getting hold of this reality on the other. Both nature,

[20] In his *System of Transcendental Idealism*, Schelling concedes that scientific discoveries may be due to the scientist being a genius, but this is not necessarily so (AA I,9,1, 323–324)—he echoes here Kant's insistence that science does not require genius. Later, however, Schelling presents Johannes Kepler as the prototype of a genius (Ziche and Rezvykh 2013, 19, 41–48). Again, the traditional link between genius and aesthetic production is still present here: Kepler's science can be characterized both in terms of scientific achievements and of aesthetic production. In any case, we here find a very early statement to the effect that scientists can, and should, be genius. On the tradition of adopting the term "genius" for activities in the realm of science, see Yeo (1988). Note, however, that typical eighteenth-century discussions of "philosophers" (the term "scientist" did not yet exist) and researchers in the field of the sciences as being "geniuses" balance the creative force of imagination with the controlling influence of careful judgement (this discussion is still present in Kant's *Critique of the Power of Judgment*). Given this background, Schelling's adoption of the label "genius" for science is only the more remarkable.

as the object of scientific investigation, and science itself are characterized by the openness of an original chaos or of genius-structured drives, combined with control structures that lead to stabilized objects or insights. Again, he uses the terminology of drives: An interesting quote in his *Erster Entwurf eines Systems der Naturphilosophie* oscillates in a complex way between a philosophical analysis of nature on the one hand, and more methodological, philosophy-of-science-like issues on the other. Nature presents us with the drama of a continuous "conflict between form and lack of form", but we also see a "drive to unite everything that is individual in nature into a joint product". This drive in itself brings this conflict under control in such a way that it determines a "certain sphere of possible forms". Here, Schelling combines a number of motives: An original conflict, an integrating function, and a determined sphere of products. This is important: He does not argue here in terms of a sequence of steps but aims at identifying a structure that is sufficiently complex to fulfil all these tasks at once. When he later talks in terms of a two-step procedure of original chaos and subsequent control operations, he further explicates this intricate structure. Note also that he summarizes this entire structure as giving the impression that a "shared ideal" had been present in the mode of a "Vorschweben", "that […] a shared ideal had soared/hovered" in front of us as a guiding ideal in nature—again, one of the standard terms of open future-directed faculties is used here (AA I,7, 92; see also 100–101).

4 The Mechanics of Drives: A Non-directional "sollicitatio ad motum"

The focus upon drives in the foundations of scientific practice introduces a conceptual problem: We need a description for states that are open, comprehensive, not singling out a particular goal, but that nevertheless are not inert; that are active without already developing themselves in a determined fashion. In drive-terms, this feature of drives is stated very explicitly in Blumenbach's writings: The fundamental drive that underlies the phenomena of generation, nutrition, and reproduction is everywhere the same; it specifies and gets specified according to the fundamental conditions in which it is set into motion (Blumenbach 1781, 19). This problem is well-known in eighteenth-century mechanics, and Schelling takes up a key term from mechanics, the term "Sollicitation". This term explicitly asks for an understanding of the origin of movement and for understanding a transition from states of rest into movement. This hints at yet another important context for Schelling's considerations, namely ideas from mystical/Pietist contexts: Jakob Böhme explicitly combines the terms "Trieb" and "Quell" (e.g. Böhme [1656], 24), "drive" and "source", and Schelling adopts the term of a "Selbstbewegungsquelle" from the writings of Friedrich Christoph Oetinger.

One of the standard images for "drives" is that of a (mechanical, hydrodynamic, etc.) *pressure* (as in Freud's hydrodynamic imagery in his theory of "libido" and of drives).[21] This image is significant. It links up with key features of the notion of "drive" as they have been developed so far: A pressure, typically, is not a linear propagation of a force or movement in a particular direction. Talking about pressure requires that a force meets resistance; typically, pressure is measured as force per surface (as in an air-filled balloon). The aspect of resistance is of key importance here, not only because this is structurally equivalent to the two-forces equilibrium as we know it from Kant's *Metaphysical foundations*, from Fichte's discussion of the subject's exploration of its environment, or from Schelling's philosophy of nature, but more importantly because it links up movement, or moving forces, with states of rest. Describing a mechanical situation in terms of pressure typically refers to states of rest, but at the same time says that the system is about to get moving. This being just on the brink of setting into motion, or being set into motion, is captured, in technical language within eighteenth-century mechanics, as a "sollicitatio ad motum".

In mathematical mechanics, "sollicitatio" is used to describe and analyse the transition from a state of rest into a state of movement, with the equilibrium state of a lever or the highest point in a vertically upward throw as paradigm cases. What these cases share, is that they refer to a state or a moment of rest that, however, is not stably inert. An upward throw gives a very precise illustration of the notion of "sollicitatio": In an

[21] Many thanks to Michael Gutfleisch for discussions about the pervasive importance of the notion of "pressure", the links between this notion and that of an equilibrium of forces, and the imagery related to this concept.

infinitesimally small moment, the object is at rest, after the movement it had in the beginning has vanished away, but directly the object is starting to move again—this can be taken to indicate that in this single moment of rest, there still is, somehow, movement, but then precisely in the form of a "sollicitatio". One way of capturing this can make use of infinitesimal considerations: A state of rest is conceptualized as a state of infinitesimally small motion, a virtual velocity.[22]

Schelling gives this term a meaning that reaches far beyond the mechanical connotations of this term (that still have been the exclusive context of this term in Kant's oeuvre). An important passage in his text *On human freedom* combines the mechanical imagery with a term ("Selbstbewegungsquelle") that he adopts from the mystical-pietist context via the writings of Friedrich Christoph Oetinger: "Man is placed on that summit where he has in himself the source of self-movement toward good or evil in equal portions" (AA I,17, 143, see also AA I,17, 261 on the relevant sources; Schelling 2006, 41; on Schelling and Oetinger see McGrath 2018), he is placed in a position and moment of (eternal?) decision, at the "threshold", a "Scheidepunkt": "That is why there must be a general ground of solicitation, of temptation to evil, even if it were only to make both principles come to life in man, that is, to make him aware of the principles" (AA I,17, 143; Schelling 2006, 41). This is an important passage in its grammatical ambiguity: The German phrase "Grund der Sollicitation" can be read as either being about (gen. subj.) "solicitation" itself being a ground or reason, or about (gen. obj.) "solicitation" requiring a ground/reason. The double grammatical function of the genitive-construction here precisely mirrors the double function of a (philosophical, metaphysical, mechanical) origin, a starting point, as a "Selbstbewegungsquelle". Also, the strongly present mechanical imagery brings these two dimensions together. A state that is described via the term "solicitation" is a state that comes about via a previous process (think again of a vertical throw) that is the reason/ground for this state's being able to be a reason/ground (see also AA I,17, 165). Schelling has a

[22] Note, however, that this image is not fully adequate because the source of motion is not in the object itself, but derives from its position in the gravitational field, and because this image cannot capture a *state* of being solicited towards movement. For the terminology in Schelling's times, see the entries in Gehler (1787–1796, vol. II, 567, vol. I, 606). On this term and the relevant context, see Bonsiepen (1995), Jammer (1957), Boudri (2002), Ziche (1996).

number of alternative descriptions for these dynamically fertile states: “Aufwallen des Centri”, “welling up in the *centrum*” (AA I,17, 144; Schelling 2006, 42),[23] an “efficient foundation”, a “wirkender Grund” (AA I,17, 144).[24]

What Schelling wants to capture with the way how he puts the notion of “drive” to use in his philosophy is a state that is both primordial and complex, both omnidirectional and non-directional, but that can be given a direction via an external stimulus. This he states explicitly in his treatise on the *World-Soul* when discussing the drives of animals; “the direction of this drive [the drive towards movement] is, originally, undetermined”; “this drive, as any drive, is originally directed towards an undetermined goal. Its direction is determined by an external stimulus” (AA I,6, 249). It is this complex internal structure of drives that allows Schelling to define the term “drive”, in the same context, as the very combination of freedom and lawfulness in nature (AA I,6, 216).

What we get here is the picture of a fully open, undetermined, chaotically unstructured initial state that in a second step becomes controlled. This, however, does not happen via imposing permanent limitations on the original state; rather, one of the many options that the original state offers becomes actualized. This is perfectly fitting for Schelling’s attempts at giving us a framework for understanding the dynamics of research: He is not interested in inertly stabilized results, but in an interplay between open and directed dynamics.

5 "Vergeistigung": New Horizons for "Drive"-Concepts

Schelling uses the notion of “drives” in contexts that integrate activity and receptivity in humans or that combine rest and motion, science and creativity. The underlying idea is that of a comprehensive openness that

[23] This translation loses the deliberate ambiguity of a centre that sets eruptive motion going and that itself wells up. Cf. again Böhme (1656, 42): “trieb und wallen Gottes”. The term “wallen” already features in the full title of Böhme’s text.

[24] See also SW I,2, 178, where Schelling combines “Sollicitiation”, “Streben”/“striving”, and “Zuckung”/“convulsion”, thus highlighting the infinitesimal, fit-like connotations of this term.

is concretized by choosing or actualizing one of the options available in the original state—in the image of a pressurized "solicitation": one of the many directions that a movement can take, in departing from the richly complex original state. He takes an important step by claiming this image also for cognitive contexts, most strikingly for the dynamics of scientific research. The importance of the step of ascribing to the scientist the status of genius (this semantic innovation requires more research; clearly, Schelling was a key figure in this conceptual innovation) cannot be overestimated. This step is part of a profound reassessment of our understanding of knowledge: Knowledge and the acquisition of knowledge need to be described in non-hierarchical ways; Schelling does not think in terms of deriving results from first principles in a deductive sequence of steps. Neither does he argue inductively, in a traditional sense: The original states that lie at the beginning of science already are inherently rich and complex; the abstractive model of classical induction does not fit here. Rather, Schelling's arguments lend themselves readily for a hermeneutic understanding in which operations of embedding in broader contexts and of an interplay between totalities and specific items stand central (see Ziche 2011 on Schelling's notion of 'construction'; McGrath 2018, 182 on Schelling's "non-representational theory of knowledge").

The link between a drive-discourse and cognitive contexts is not restricted to Schelling. Reimarus' "Vorstellungstriebe" have already been mentioned; Franz von Baader explicitly identifies an "Erkenntnisstrieb", a "cognitive drive" that "coincides" with the "Gestaltungs- und lebendigen und organischen Bildungstrieb", with "a formative, living, organic drive", insofar as all of these drives are everywhere directed towards "procreation, giving birth, pronouncing and representing a word, name, image".[25] Human drives are human; other than the animals' "Kunsttriebe", they are not fixed, not restricted to stereotypically repeated productions. But they still are drives: While being human in their openness, they remain drives in their being beyond rational control. This is characteristic both for the production of artworks, as discussed in the final chapter of

[25] Baader (1808, 117); he also uses the term "Sollizitation" (Baader 1808, 114). Interestingly, in comparing cognitive and sexual drives, he emphasized the hermaphroditic character of cognitive drives (Baader 1808, 118–119), thus again arriving at an integrative form of drive. See also Ffytche (2012, 131).

Schelling's *System of Transcendental Idealism*, and—a point that Schelling comes to realize more and more as his thinking advances—in the practice of scientific research. For understanding the course of science, we need to go beyond knowledge; in its beginning and in its (unattainable) final state, science can no longer be explained in terms of knowledge, or of hierarchical explanations. Knowledge needs to step outside of a linear progression; each point on the path of science or of acquiring knowledge more generally needs to be considered a critical point that requires both reflection and strong decisions. Put differently: Each point on this course should be a moment of "sollicitation". The picture of science that we then get is, perhaps surprisingly, rather sober and straightforward: The openness of drives prevents us from hasty, one-sided, prejudiced thinking, and at the same time, acting upon drives and instincts keeps the process of acquiring knowledge going and makes sure that this remains a process that is rife with risks and surprises, everywhere.

In an emblematic passage in his dialogue *Clara*, that is, in a context in which mystical and finitude-transcending ideas feature prominently, Schelling describes an operation that transforms knowledge into something more subtle, evasive, intangible, but at the same time omnipresent; or, in different terminology: he describes how knowledge is both made more subtle, how it evaporates, which at the same is an intensification of knowledge and only leads deeper into the dimension of the mind. The key term here is that of "Vergeistigung",[26] of "spiritualization", of becoming more and more subtle but in this process becoming more and more mind-like (and in doing so, the future-directed but non-deductively determining faculties, here that of "Ahndung", become important). We might say: In this process, knowledge is properly turned into a drive while, at the same time, only becoming the more essentially mind-like:

> It was not enough, one said, to have given up the connection to what is objective or to insensible nature, if so coarse a concept as that of knowledge were still tolerated within the subjective. Knowledge itself was still too solid; spiritualization would be perfect only when, instead, a tender, fleet-

[26] See also AA I,11,1, 354, where Schelling renders the classical topos of the poets' being driven by divine inspirations in drive-terms; see also SW I,7, 49.

ing spoor of a feeling or hunch ["zarter, flüchtiger Duft von Ahndung und Gefühl"; the connotations of these metaphors are somewhat lost in the translation] alone remains; that is, when the subjective is subjectified again. (SW I,9, 4; Schelling 2002, 3–4)

References

Baader, Franz von. 1797. *Beyträge zur Elementar-Phisiologie*. Hamburg: Carl Ernst Bohn.

———. 1808. Ueber die Analogie des Erkenntniss- und des Zeugungs-Triebes. *Jahrbücher der Medicin als Wissenschaft* 3 (1): 113–124.

———. 1809. Ueber die Behauptung: daß kein übler Gebrauch der Vernunft seyn kann. In Franz von Baader: *Beiträge zur dinamischen Philosophie im Gegensaze der mechanischen*, 120–124. Berlin: Realschulbuchhandlung.

Barentsen, Gord. 2015. Silent Partnerships: Schelling, Jung, and the Romantic Metasubject. *Symposium* 19 (1): 67–79.

Barnouw, Jeffrey. 1972. 'Der Trieb, bestimmt zu werden'. Hölderlin, Schiller und Schelling als Antwort auf Fichte. *Deutsche Vierteljahrsschrift für Literaturwissenschaft und Geistesgeschichte* 46: 248–293.

Berg, Robert. Jan. 2003. *Objektiver Idealismus und Voluntarismus in der Metaphysik Schellings und Schopenhauers*. Würzburg: Königshausen & Neumann.

Blumenbach, Johann Friedrich. 1781. *Über den Bildungstrieb und das Zeugungsgeschäfte*. Göttingen: Johann Christian Dieterich.

Böhme, Jakob. 2009 [1656]. *Morgenröte im Aufgangk*. In Jakob Böhme: *Werke. Morgenröte im Aufgangk. De signatura rerum*, ed. Ferdinand van Ingen, 9–506. Frankfurt am Main: Deutscher Klassiker Verlag.

Bonsiepen, Wolfgang. 1995. Entry "sollicitatio". In *Historisches Wörterbuch der Philosophie*, vol. 9, 1056–1058. Basel: Schwabe.

Boudri, J. Christiaan. 2002. *What was Mechanical about Mechanics. The Concept of Force Between Metaphysics and Mechanics from Newton to Lagrange*. Dordrecht: Springer.

Cesa, Claudio. 2002. Sensibilité et conscience. Remarques sur la théorie des Triebe chez Fichte. *Revue germanique internationale* 18: 121–132.

Durner, Manfred, Francesco Moiso, and Jörg Jantzen. 1994. *Wissenschaftshistorischer Bericht zu Schellings naturphilosophischen Schriften*

1797–1800. Supplementary volume to AA I,5–9. Stuttgart-Bad Cannstatt: frommann-holzboog.

Ffytche, Matt. 2012. *The Foundation of the Unconscious. Schelling, Freud and the Birth of the Modern Psyche*. Cambridge: Cambridge University Press.

Fichte, Johann Gottlieb. 1969 [1794/1795]. *Grundlage der gesammten Wissenschaftslehre*. In *Johann Gottlieb Fichte-Gesamtausgabe*, vol. I,2. Stuttgart-Bad Cannstatt: frommann-holzboog.

Gehler, Johann Samuel Traugott. 1787–1796. *Physikalisches Wörterbuch oder Versuch einer Erklärung der vornehmsten Begriffe und Kunstwörter der Naturlehre mit kurzen Nachrichten von der Geschichte der Erfindungen und Beschreibungen der Werkzeuge begleitet in alphabetischer Ordnung*. Vol. 6. Leipzig: Schwickert.

Grimm, Jacob, and Grimm, Wilhelm. 1854–1961. *Deutsches Wörterbuch*. http://woerterbuchnetz.de/cgi-bin/WBNetz/wbgui_py?sigle=DWB.

Hagen, Theodor Alexander von. 1808. *Kosmologische Geschichte der Natur, insbesondere des Mineral- und Pflanzenreichs der Erde*. Heidelberg: Schwan & Goetz.

Jacobs, Wilhelm G. 1967. *Trieb als sittliches Phänomen. Eine Untersuchung zur Grundlegung der Philosophie nach Kant und Fichte*. Bonn: Bouvier.

Jammer, Max. 1957. *Concepts of Force. A Study in the Foundations of Dynamics*. New York: Dover.

Kant, Immanuel. 1913 [1790]. *Kritik der Urteilskraft*. In Immanuel Kant: *Gesammelte Schriften (Akademie-Ausgabe)*. Vol. 5. Berlin: Reimer.

———. 1917 [1800]. *Anthropologie in pragmatischer Hinsicht*. 2nd ed. In Immanuel Kant: *Gesammelte Schriften (Akademie-Ausgabe)*. Vol. 7. Berlin: Reimer.

———. 2000. *Critique of the Power of Judgment*, ed. Paul Guyer. Cambridge: Cambridge University Press.

———. 2007. *Anthropology, History, and Education*, ed. Robert B. Louden and Günter Zöller. Cambridge: Cambridge University Press.

König, Christian. 2016. *Unendlich gebildet. Schleiermachers kritischer Religionsbegriff und seine inklusivistische Religionstheologie anhand der Erstauflage der Reden*. Tübingen: Mohr Siebeck.

Lind, Gunter. 1992. *Physik im Lehrbuch 1700–1850: Zur Geschichte der Physik und ihrer Didaktik in Deutschland*. Berlin: Springer.

McGrath, Sean J. 2010. Schelling on the Unconscious. *Research in Phenomenology* 40: 72–91.

———. 2018. Retrieving the Schellingian Tradition. Friedrich Christoph Oetinger's Speculative Pietism. *Kabiri. The Official Journal of the North American Schelling Society* 1: 175–192.

Oken, Lorenz. 1843. *Lehrbuch der Naturphilosophie*. 3rd ed. Zürich: Friedrich Schultheß.

Pénisson, Pierre. 2002. Trieb et énergie chez Herder. *Revue germanique internationale* 18: 45–52.

Reimarus, Hermann Samuel. 1790. *Allgemeine Betrachtungen über die Triebe der Thiere, hauptsächlich über ihre Kunsttriebe. Zur Erkenntniß des Zusammenhanges der Welt, des Schöpfers, und unser selbst*, letzte verbesserte Auflage, vol. 1. Wien: F. A. Schrämbl.

Richards, Robert J. 2002. *The Romantic Conception of Life: Science and Philosophy in the Age of Goethe. Science and Its Conceptual Foundations*. Chicago: University of Chicago Press.

Schelling, Friedrich Wilhelm Joseph. 1966. *On University Studies*, trans. E.S. Morgan and ed. Norbert Gutermann. Athens, OH: Ohio University Press.

———. 1978. *System of Transcendental Idealism (1800)*, trans. Peter Heath. Charlottesville: University Press of Virginia.

———. 1994. *On the History of Modern Philosophy*, ed. Andrew Bowie. Cambridge: Cambridge University Press.

———. 2002. *Clara or, On Nature's Connection to the Spirit World*, trans. Fiona Steinkamp. Albany, NY: State University of New York Press.

———. 2006. *Philosophical Investigations Into the Essence of Human Freedom*, trans. Jeff Love and Johannes Schmidt. Albany, NY: State University of New York Press.

Schiller, Friedrich. 1966 [1801]. *Über die ästhetische Erziehung des Menschen in einer Reihe von Briefen*. In Friedrich Schiller: *Werke in drei Bänden*, ed. Herbert G. Göpfert, vol. 2, 445–520. München: Hanser.

Schleiermacher, Friedrich. [1799]. *Über die Religion. Reden an die Gebildeten unter ihren Verächtern*. In Friedrich Schleiermacher: *Kritische Gesamtausgabe*, vol. I,2, 185–326. Berlin and New York: de Gruyter 1984.

Stiening, Gideon. 2002. Entre Fichte et Schiller. La notion de *Trieb* dans le *Hyperion* de Hölderlin. *Revue germanique internationale* 18: 87–103.

Yeo, Richard. 1988. Genius, Method, and Morality: Images of Newton in Britain, 1760–1860. *Science in Context* 2: 257–284.

Ziche, Paul. 1996. *Mathematische und naturwissenschaftliche Modelle in der Philosophie Schellings und Hegels*. Stuttgart-Bad Cannstatt: frommann-holzboog.

———. 2011. Das System als Medium. Mediales Aufweisen und deduktives Ableiten bei Schelling. In *System und Systemkritik um 1800*, ed. Christian Danz and Jürgen Stolzenberg, 147–168. Hamburg: Felix Meiner.

———. 2016. Epistemic Confidence—Kant's Rationalization of the Principles of Seeking and Finding. *Internationales Jahrbuch des Deutschen Idealismus/International Yearbook of German Idealism* 14: 81–102.

Ziche, Paul, and Peter Rezvykh. 2013. *Sygkepleriazein. Schelling und die Kepler-Rezeption im 19. Jahrhundert*. Stuttgart-Bad Cannstatt: frommann-holzboog.

14

The *Trieb* of Dialectic: Systematic and Thematic Extension of the Concept of *Trieb* in Hegel

Angelica Nuzzo

Hegel's philosophy often presents itself, for better or for worse, as a systematic *summa* of the philosophical tradition as it purports to encompass a vast range of previous historical positions with the well-known embrace of a dialectical *Aufhebung*—Hegel's signature critical gesture which is at once negating and preserving. The interpreters, on their part, often find themselves endorsing the same stance and, just as often, criticizing its alleged hubris. There is a sense in which one can look in the same light at the terminological presence of the concept of *Trieb* in Hegel's mature philosophy, which is paradigmatically consigned to the development of the *Encyclopedia of the Philosophical Sciences* (Enz.) (1817, 1827, 1830). Indeed, Hegel's use of the term seems to inherit all the facets of the concept found in the German philosophical tradition, and, in particular, it seems to aptly sum up the contemporary and immediately preceding discussion. But Hegel's employment of the term is also, at the same time, cognizant of its conceptual root—albeit in a different linguistic and

A. Nuzzo (✉)
Brooklyn College, CUNY, New York, NY, USA
e-mail: ANuzzo@gc.cuny.edu

M. Kisner, J. Noller (eds.), *The Concept of Drive in Classical German Philosophy*,
https://doi.org/10.1007/978-3-030-84160-7_14

historical context—in the ancient tradition (first and foremost in Plato and Aristotle) of which aspects of his own philosophy are sometimes announced as direct successors.[1] However, while Hegel does refer to the contemporary discussion in appealing to the concept of *Trieb* in the philosophy of nature and spirit—from pre-Kantian Enlightenment to Kant, Fichte, and Schelling (with mentions of *Bildungstrieb*, *Kunsttrieb*, *Naturtrieb*: Enz. §365 Zusatz, Hegel 1969—1971, TW 9, 494)—he seemingly does so without problematizing the concept too much—as if the term had by then been absorbed into the philosophical vocabulary and did not pose, for him, particular issues of its own. *Trieb* does not appear to be a *terminus technicus* for Hegel. It is employed, instead, in different spheres throughout his comprehensive system of philosophy—from the philosophy of nature, where it specifically applies to the phenomenon of life and articulates the relation of the animal organism to its natural environment, to the philosophy of spirit, where it describes anthropological and psychological aspects of spirit's incipient and developing subjectivity, and where it addresses spirit's action in the realms of morality, ethical life, and world history, that is, in the constitution of spirit's objectivity. Across all these spheres, *Trieb* is a concept that most generally bridges, or rather dialectically mediates and reconciles, as it were, the contradiction of nature and spirit, necessity and freedom, subjectivity and the objective world, the inner and the outer dimensions of the described movements (of life, subjectivity, action, history) progressively integrating one into the other. In this general employment, Hegel capitalizes on the term's manifold connections to other concepts such as power, instinct, desire, need, will (in recurring dyads such as *Trieb und Bedürfnis*, *Trieb und Mangel*, and *Trieb und Neigung*). (These connections are also relevant when at stake is the gesture of importing ancient Greek terminology into German.) Finally, at this general level, it should be noted that what makes the concept particularly appealing to Hegel's dialectic-speculative philosophy is *Trieb*'s reference to the *dynamic* of a process in which the fixed substantiality of an alleged presupposed "substrate" is entirely resolved into the tension of its actual (even though only

[1] See, famously, the case of the philosophy of spirit in relation to Aristotle.

incipient) movement and action,[2] and in which inert or merely potential faculties or powers of the mind are entirely resolved in the actuality of the action they perform (as they perform it).[3]

While this initial picture already warrants that attention should be paid to Hegel's use of the concept of *Trieb*, things are more complicated—and interesting—than this first terminological and thematic overview may lead one to assume. In fact, while not a *terminus technicus* and while not taking issue (at least apparently and directly) with any contemporary discussion around the topic, Hegel's concept of *Trieb* displays a novel and unprecedented meaning on the basis of the new (and indeed surprising) systematic context in which it is introduced.

As is well known, Hegel's philosophical system comprises three main spheres (or philosophical disciplines): the logic, the philosophy of nature, and the philosophy of spirit. The Logic is the "foundation" of the following systematic spheres—where the relation of foundation is to be understood not linearly but circularly, according to the famous structure of the "circle of circles" (Hegel 1969–1971, TW 6, 571).[4] The Logic is the first and the last of the systematic disciplines in which philosophical thinking articulates its successive forms. The controversial relationship between the Logic and the successive forms of nature and spirit (and the philosophical knowledge thereof) is well captured by Hegel's image of the relation between the immanent, supporting structure of the skeleton and the full-fledged materiality that constitutes the "real" creature—this being, alternatively, a specifically natural or spiritual creature (Hegel 1969–1971, TW 6, 257). Hegel's dialectic-speculative Logic presents the structures of "pure thinking" considered in its merely logical or formal activity. Within the spheres of nature and spirit that supporting logical skeleton receives its specific conformation. Crucial for my purposes here are only two very general points. For one thing, logical thinking is the pure activity of thinking taken without a subject that thinks (such as a *res cogitans*, an *ego cogito* or a transcendental "I think") and without intentional "real" objects that are being thought. For another, Hegel's Logic is the immanent presentation (or rather

[2] Think here of Hegel's programmatic transformation of "substance" into "subject".

[3] See Nuzzo (2018a).

[4] All translations from Hegel are my own.

self-presentation) of a *dynamic process*—the dynamic process of thinking's self-determination. It is not the institution of a fixed table of categories such as the ones found in Aristotle's or Kant's logic. The task of Hegel's Logic is the task of generating the immanent movement of pure thinking in the absence of concrete or real (thinking) subjects and (thought) objects. Hegel assigns this task to the logical "method", which is that which in his stated view makes such a logic distinctively "dialectic-speculative". Now, while it is no surprise that the concept of *Trieb* appears in Hegel's philosophical account of specific forms of nature and spirit (where it describes distinctive features of the activity of natural and spiritual subjects and objects), it should come instead as a surprise to find the term employed within the Logic as characterizing pure logical forms and activities which as such are (or must be) programmatically devoid of all "real" content. (In effect, interpreters have been so puzzled by Hegel's employment of this and similar terms in the Logic as to deny that this foundational part of the system is truly a "logic"). What can *Trieb* mean when it is used to indicate the "*Trieb* of the *Begriff*" or the "*Trieb* of the Idea"? What is a logical drive? And reciprocally, what kind of "concept" or "idea" is a concept or an idea that is endowed with a *Trieb*? (Indeed, in what sense can a concept—and properly, for Hegel, *the* Concept—have or even be *Trieb*?) Significantly, it is in connection with occurrences such as these that *Trieb* plays a crucial role as *Trieb* of the "absolute idea" in Hegel's presentation of the method at the end of the Logic. And here the question to be addressed regards the nature of the *Trieb* of the method or even, I suggest, the meaning of *Trieb as logical method.*

Hegel recognizes the fundamentally *dialectical* logic of *Trieb* in natural and spiritual processes. This recognition prompts a twofold operation. On the one hand, it leads him to distil the purely logical validity or meaning of the concept thereby making of *Trieb* a decisive factor in the engine of dialectical logic. *Trieb* is that which immanently pushes forward or drives on, as it were, the process of pure thinking in its inner dialectical movement. Herein *Trieb* is the pure energy of dialectical thinking as such, not something inhering in a determinate subject or mental power, or directed to a specific object. On the other hand, this recognition allows Hegel to organize the progressive concrete or "real" occurrences of *Trieb* within the different spheres of nature and spirit presenting them as instances or embodiments of the same underlying logical *Trieb*, thereby lending to these occurrences a meaning that is systematically different from what is found in the tradition. Herein *Trieb* is qualified by the specific subjects and objects that respectively display it. But the term also captures the systematic of an unfolding movement that traverses nature and spirit displaying an underlying continuity of meaning but also crucial breaks and discontinuities.

This, then, is the constellation I address in this chapter. In the first part, I discuss Hegel's employment of the notion of *Trieb* in the Logic addressing, in particular, its emergence in the conclusive chapter on the "absolute method". I show how *Trieb* becomes, for Hegel, a crucial feature of the dialectic-speculative method that articulates the forms of pure thinking within the Logic, and then further develops the immanent succession of the "real" forms of nature and spirit in the *Realphilosophie*. Logically, *Trieb* is a fundamental structure of Hegel's dialectic *as method*. In particular, *Trieb* is what I call the logical "figure" of dialectical contradiction. This is the unique and innovative meaning that the concept receives in Hegel. On the basis of this discussion, in the second part of the chapter I show by way of two brief examples how Hegel's employment of the term to describe natural and spiritual phenomena while betraying the apparently unproblematic inheritance of contemporary and past uses is, in fact, deeply innovative. As character of animal nature first, and then of practical spirit developed on the basis of the logical method, *Trieb* captures the distinctively *dialectical* cypher of natural and spiritual phenomena, the way in which nature and spirit work their way through real contradictions.

1 *Trieb* as the Engine of Hegel's Logical Method

1.1 In the Beginning Is *Trieb*

In the conclusive chapter of the Logic dedicated to the "absolute idea", Hegel clarifies, first, that "method" is the *movement* of self-realization of the concept; and second, that this movement is the *universal and absolute activity* that pervades everything as condition of its meaningfulness and truth.[5] Method, Hegel contends, is "soul and substance" since everything true and meaningful owes its truth and meaning—hence its very subsistence in the order of rationality—to its being part of the overall logical action which has been developed up to that point. But once we get to the presentation of the method, that is, at the end of the logic, the action is *complete*. That all foregoing determinations of thinking are part of the comprehensive movement of the logic is, at this point, a demonstrated and accomplished fact. For, having the entire course of the logic behind, of every particular determination we can say that it has been exhibited in its specific place and modality within the course of the logic, and for this reason it has now become part of the logical whole—hence part of the method or of the universal "movement of the concept" (Hegel 1969–1971, TW 6, 551). This is indeed the condition of truth and meaning for all particular determinations, namely, to be part of the broader and more comprehensive movement of pure thinking. Conversely, at this point Hegel maintains that the idea "is the highest power (*Kraft*) or properly the only and absolute power of reason (*die einzige und absolute Kraft der Vernunft*)"—and then he adds, "not only that, but it is also the only and highest *Trieb* [of reason] to find itself in everything" through its own action (Hegel 1969–1971, TW 6, 551–552). Thus, the absolute idea, which is "absolute method" and has just been presented as the "soul" of the logical movement, is not only *Kraft* but also *Trieb*. The two are not identical. Thus, the method's universality, that is, the universal validity of the movement enacted in the logic by pure thinking is the universality of the *Trieb* of reason to affirm and recognize itself as an omni-pervasive force. Ultimately and retrospectively, it is now revealed that precisely this *Trieb* and its underlying *Kraft* is responsible for having weaved the entire development of the logic. On this basis, that is, on the basis of what it has accomplished (in its power or *Kraft*), reason's *Trieb* is now presented as "method". Method is not just that which has immanently produced the logical movement so far as the full actualization of reason's power. In addition, at this conclusive stage, method is reason's—or rather the idea's—"only and highest *Trieb*" to *recognize* itself in all that it has

[5] For a comprehensive account of the method, see Nuzzo (2018b).

exhibited within the logical movement, the endeavour and drive to reclaim that movement as its own. This, however, is also the springboard that systematically leads beyond the logic to the development of the philosophy of nature and spirit. Herein at stake is, once again, the idea's *Trieb* to find itself in all the different forms of natural and spiritual actuality—the idea's *Trieb* to find itself in its other.

In good Aristotelian fashion, the method displays, for Hegel, three moments, beginning, advancing, and ending. At stake are three distinctive actions of pure thinking in generating three moments of the logical process. Hegel begins with the beginning.[6] What does it mean to raise the question of the beginning as a *methodological* question? What is the beginning "in and for itself", as "mode" [*Art und Weise*] of movement [independently of what/who it is that begins]? How does the structure of beginning inform the process staged by the Logic, the movement of the concept *as movement*? Significantly, these questions now replace metaphysical (theological and cosmological) questions of *origin*. In the difference that separates the question of origin from the issue of what is, methodologically, the action of beginning lies the difference between traditional metaphysics (and Kant's critique thereof) and Hegel's dialectic-speculative logic.[7]

"A beginning (*arche*) is that which does not itself follow necessarily from something else, but after which a further event or process naturally occurs" (Aristoteles 1831, *Poetica*, 7, 1450b, 25–27). The beginning is original (first) *Trieb*. *To begin* (as intransitive or absolute action) is the action characterized by being an "immediate" that has the form of "abstract universality" (Hegel 1969–1971, TW 6, 553). As moment of the method, "[T]he beginning has no other determinateness than this: being simple and abstract" (Hegel 1969–1971, TW 6, 554), immediate and universal. And yet, as beginning *of a dynamic process*, this immediate and simple is most essentially the *Trieb* that generates the further movement. At the level of the method, Hegel is concerned with the way in which the beginning is made (whatever the beginning in its content-determination may be) in order for the logical movement, and eventually

[6] TW 6, 553: "Es ist dabei *erstens* mit dem *Anfange* anzufangen".

[7] I discuss this point extensively in Nuzzo (2012, chapter 4).

for the end of the entire movement, to immanently issue. Methodologically, the beginning is a *dynamic* beginning, which accordingly can be defined as *Trieb*—as the concept's *Trieb* for a further advancement (Hegel 1969–1971, TW 6, 555), as the concept's *Trieb* to realize itself (Hegel 1969–1971, TW 6, 554). In other words, the method discloses that the beginning of discursive thinking's activity is the dynamic tension of a logical *Trieb*. To appreciate the importance of Hegel's position here, one should recall the famous polemic against Schelling in the preface to the *Phenomenology of Spirit*. The beginning of philosophy—and of philosophical thinking—cannot be the Absolute. It cannot be an ontological absolute but, for that matter, any kind of absolute—and certainly not the absolute of intellectual intuition. In fact, Hegel's point is that in the Absolute there is (there cannot be) any *Trieb* to further realization. Now, at the end of the Logic, Hegel tells us what replaces, methodologically, the failed beginning with the Absolute. It is a pure logical *Trieb*. Most originally, *Trieb* is the non-absolute of discursive thinking: it is that which is justified as a beginning only in the act of moving on away from itself; it is that which truly is what it actually is only in the tension or drive that makes of it something else, namely, the advancement—the second moment of the method. On this basis, the *Trieb* of method (or of the idea) is the dynamic exhibition of dialectical contradiction but also displays an inner teleological structure without assuming the concept of a purpose (*Zweck*). The *Trieb* of method, to borrow Kant's terminology, exhibits the inner "purposiveness without purpose" that drives the logical movement on. It is the dynamic tension that ultimately leads the beginning to the end—to the Logic's conclusion.

Hegel notices that non-dialectical thinking (what he calls "external reflection"), while embracing the claim put forth by dialectical-speculative thinking that the beginning is simple and abstract, accepts such beginning only for the sake of a promised content. Thereby, the beginning is transformed into an arbitrary assumption (or into a provisional or merely hypothetical beginning *à la* Reinhold: Hegel 1969–1971, TW 6, 570) made only in order to satisfy thinking's desire (*Streben*) to move on away from it. Furthermore, "mere opinion" in its *Bewußtlosigkeit* defines the "simple" and "abstract" character of the beginning as a content that is actually given ("*es gibt*": Hegel 1969–1971, TW 6, 555)—either in reality or in thinking. On the contrary, the method, which is the "consciousness of the concept" (Hegel 1969–1971, TW 6, 555), understands the simplicity and abstractness of the beginning as its mere formality, as the "objective, immanent form" that is "in itself" "lacking (*mangelhaft*)" and endowed with the *Trieb* to realize the concept. In other words, the illusion of *Streben* affecting external reflection, which seems to make the beginning alternatively provisional, instrumental, or irrelevant, is overcome once we recognize that such *Trieb* is nothing else but the immanent form of the beginning taken as moment of the method. The connection *Trieb*/*Mangel* is relevant here. The beginning is necessarily "lacking" because it is *only* the beginning (of the overall movement; of an advancement), hence necessarily incomplete. To begin is necessary in order for thinking to be set in motion. Yet the beginning taken in isolation is not yet a sufficient characterization of movement but only its inception. It is this predicament of dialectical thinking that Hegel captures as he attributes *Trieb* to the logical concept and idea. To consider the beginning as moment of the method then is to gain the consciousness of the fundamental incompleteness of all beginning action as such, of its need to be followed by an advancement which completes it and realizes it; but, most properly, it is to gain consciousness of the fact that the beginning itself is immanently endowed with the "impulse and drive"—*Trieb*—"to further develop and determine itself" (Hegel 1969–1971, TW 6, 555). This is the universality of the method. It is precisely in this way that the beginning leads on to the second moment of method, namely, the "advancement (*Fortgang*)" (Hegel 1969–1971, TW 6, 555f.). It should be noted that against a too simple external teleology, *Trieb* and *Streben*, for Hegel, indicate the formal character of the beginning not the character of the advancement. In the beginning is contained the seed or indeed the impetus of the entire movement of the whole; and yet, since the beginning is inherently "lacking", it is also true that the development is not (analytically) contained in the beginning, only its *Trieb* is. Every moment of the process of the method, Hegel confirms, is synthetic and analytic at the same time (Hegel 1969–1971, TW 6, 557).

Thereby the logical method offers an important lesson when at stake, this time, are determinately "real" processes in nature and spirit—in organic life, in the soul, in spirit's moral and ethical action, in world history. *Trieb* animates all these processes—and is present in all these spheres—insofar as they are only incipient, still immediate, and merely abstract: *mangelhaft*, as it were. But it is present in them to the extent that an actual true beginning is made, that is, to the extent that the seed for a further advancement or development is planted (consider, for instance, the "beginning" of spirit in the natural *Trieb* of the animal organism or the *Trieb* which is the incipient, utterly abstract and immediate form of the will). Moreover, the presence of *Trieb* in real processes discloses both the necessity of the advancement resulting from the drive, and the range of possible *contingent* manifestations that issue from the drive (e.g. think of moral action or of historical processes). This is indeed an implication of the logic of *Trieb*. For, as the method puts it, the connection beginning/advancement displayed by the logical *Trieb* is both analytic and synthetic.

1.2 The Logical *Trieb* of Contradiction

Before getting to the implications of Hegel's logical and methodological account of *Trieb* in select moments of the philosophy of nature and spirit, I need to complete that account by briefly looking at its relation to the logical form of dialectical contradiction. The point has emerged already in the previous discussion. The beginning as moment of method is *Trieb* to the extent that herein the idea's action is and is not what it essentially is. The beginning is as such a beginning but precisely insofar as it is *only* a beginning it is first fulfilled in the advancement, that is, in what the beginning properly is *not*. In the Logic of Essence, dealing thematically with "contradiction" as the third of the "determinations of reflection" (*Reflexionbestimmungen*) after "identity" and "difference", Hegel counters the common positions that deny the real occurrence or the occurrence in existence of contradiction. Ordinary thinking removes contradiction from real things and limits it to a formal logic—and a formal truth—that allegedly has nothing to say about the world (this is an extension of what Hegel, referring to Kant, ironically calls the "*Zärtlichkeit für die weltlichen Dinge*": Enz. §48 Anm.). As the saying goes, so Hegel, "contradiction *is* not ([der] Widerspruch nicht *gebe*)". Properly, for common non-dialectical thinking contradiction "can neither be represented nor thought". From which it follows that "[w]hether it occurs in actuality or in thought's reflection, contradiction is universally reckoned as an accident, an abnormality, as it were, a momentary fit of sickness" (Hegel 1969–1971, TW 6, 75). Hegel counters this conclusion by appealing, on the one hand, to what his Logic has exhibited up to that point (namely, determinations of Being such as finitude and the infinite, and Essence), and by appealing on the other hand to our "common experience", which easily detects "many contradictory *things*" or that contradiction, in fact, *is* (Hegel 1969–1971, TW 6, 75—my emphasis). Hegel's crucial claim, however, is that contradiction is "the negative in its essential determination, the principle of all self-movement which consists in nothing else than in the display of contradiction". Endorsing the background of Zeno's famous paradoxes (and of the "old dialecticians" more generally), Hegel repeats that the first, immediate proof of the *Dasein* of contradiction is found in "external sensible movement" (Hegel 1969–1971, TW 6, 76). But the next crucial example or figure of contradiction in existence is "*Trieb überhaupt*". *Trieb* is here introduced in its purely *logical* determination as the fundamental and distinctive dynamic character of "*Lebendigkeit*", whether it is the vitality or liveliness that animates the animal organism or animal subjectivity (Enz. §§350, 351) or the vitality proper to spiritual subjectivity in all its forms. Moreover, in its logical validity as figure of contradiction *Trieb* here sums up a range of historical positions: "appetite, or the nisus of the monad, the entelechy of the absolutely simple essence", Hegel spells out with references to Leibniz and Aristotle as examples. In contrast but also in continuity with external movement, "drive in general" is "internal self-movement, self-movement proper". *Trieb* is the contradiction in existence and in action. It expresses "nothing else than that something is, in itself, itself and the lack (*Mangel*) of itself (the negative), in one and the same respect. Abstract self-identity is not yet vitality (*Lebendigkeit*); but the positive, since implicitly it is negativity, goes out of itself and sets its alteration in motion. Something is alive, therefore, only to the extent that it contains contradiction within itself: indeed, force (*Kraft*) is this, to hold and endure contradiction within" (Hegel 1969–1971, TW 6, 76). *Trieb* and *Kraft* are logical figures of dialectical contradiction. As the inner movement that goes outside of itself, *Trieb*, Hegel poignantly claims, is "der *gefühlte* Widerspruch" (Enz. §204 Anm.), the felt contradiction that unlike the abstraction of a dead and unmoved self-identity generates the dynamic of self-movement and self-transformation. This is the same logic that leads Hegel to attribute *Trieb* and *Kraft* to the absolute idea as "absolute method". But it is also the logic that underlies the thematic introduction of the animal drive in the philosophy of nature and of practical drives in the philosophy of spirit. In these realms, however, self-movement and self-transformation, and the *Trieb* and contradiction responsible for them, will present, respectively, an additional specific phenomenology.

2 The Reality of *Trieb* in Nature and Spirit

We have seen that, at the end of the Logic, Hegel claims that the idea—and speculative reason—is endowed with "the only and highest *Trieb* to find itself in everything" (Hegel 1969–1971, TW 6, 551–552; see Enz. §376 Zusatz, Hegel 1969–1971, TW 9, 539). This *Trieb* becomes the systematic basis for the development of the *Realphilosophie*, that is, for the account of the ways in which the idea—or rationality—is displayed and finds and knows itself in the realms of nature and spirit. Methodologically, Hegel's philosophy of nature and spirit are the product of the idea's logical *Trieb*. This, I suggest, is precisely what distinguishes Hegel's systematic account of the specific reality of *Trieb* in these spheres from all his contemporaries.

In what follows I concentrate on two moments of the development of the concept of drive in Hegel's *Realphilosophie*, namely, the animal drive in the last stage of the philosophy of nature and the drive thematized in the Psychology of subjective spirit. I shall read these two stages comparatively, and present Hegel's immanent development of the real figure of *Trieb* across these spheres as the dialectical juncture of nature and spirit. Indeed, in the development of this concept we can assess both the way in which spirit is "anticipated" in nature and the way in which nature is maintained and yet transformed within spirit's liberation from nature. In this discussion, I shall focus on the points of continuity between the logical account of *Trieb* and its concrete real manifestations.

Both in organic nature and in subjective spirit *Trieb* indicates the *practical* attitude of the individual towards the environment and towards itself, an attitude that is both affirming and negating. At the level of animal subjectivity, *Trieb* expresses the animal's still merely *reactive* tendency towards the external, inorganic world. Herein *Trieb* is a capacity for action that reaffirms the unity of the species, not the individuality of the animal individual, which is ultimately negated in death. In the sphere of spirit, by contrast, *Trieb* acquires the creative and transformative features that characterize spirit's distinctively self-conscious and free subjectivity beyond nature's mere givenness and beyond the animal's merely re-active tendency. In general, the subject's action, which spans the development from animal to human subjectivity, presents two possible relationships, namely, the theoretical, in which the individual, reflected in itself, affirms itself in a negative relation to the world; and the "practical relation", which seems, at least initially, to have the opposite direction whereby the external world is viewed as fulfilling the negation or the lack perceived within the subject. In the practical relation, the "real process" of the organism begins with "the diremtion (*Diremtion*) in itself, with the feeling of exteriority as the *negation* of the subject", which, however, "is also, at the same time, the positive relation to itself and its *certainty* in front of such negation". This latter, as the capacity of turning a negative external influence into a positive self-certainty, is captured by the notion of *Trieb*. The practical relation of the organism to inorganic nature is "the feeling of *lack* (*Mangel*) and the *Trieb* to overcome it" (Enz. §359). Herein, following the logic whereby *Trieb* is integral to the felt *Mangel*, the drive embodies the way in which dialectical double negation works in the animal organism thereby becoming, most importantly, the distinctive cypher of life. Indeed, Hegel declares, "*[n]ur ein Lebendiges fühlt Mangel*" (only a living being feels a lack/want) and consequently has the impulse to overcome it thereby affirming itself as subject. Significantly, for Hegel, the basis of this claim is the logical position discussed earlier. For, he maintains, "only the living being is, in nature, the *concept*, which is the unity of itself and of its determinate opposite" (Enz. §359 Anm.). The living being is the only true instance of the "concept" in nature; it is the "concept that enters existence". And the concept, we have seen in the last chapter of the Logic, as dialectic-speculative concept, is endowed with the (logical) *Trieb* to realize itself in the unity of itself and its opposite (which is the Idea). Hegel notices that while a mere "limit" is a negative only for a third external term, not for the one that is affected by it, "*Mangel*" is negation insofar as it already implies the "*Darüberhinaussein*"—the being-beyond-it—that is properly *Trieb* and that betrays the fact that "contradiction is immanent and posited" in the being who *feels* the lack and only in this being. Now Hegel maintains that only a being who bears "the contradiction with itself within itself" and, most significantly, "is able to *endure* it", only this being can be said "subject". The connection between "lack" and "drive" constitutes here the "infinity" of the subject (Enz. §359 Anm.). The action of *Trieb* dialectically resolves the subject's finitude or limitation into its infinity. While the "lack" is particularized as "need (*Bedürfnis*)", the "drive" expresses the "activity to overcome the lack" in its content-determination. More radically, the drive overcomes the very "form" of the need, that is, the fact that need is "only a subjective". In the movement of *Trieb*, the content as need becomes the "purpose (*Zweck*)" of the drive's activity, while the drive becomes "instinct" (Enz. §360).

At this point, Hegel adds a remark that is significant when at stake is the interpretive task of measuring his distance from the contemporary debate. He notices that it is precisely the dialectical connection between *Mangel*, *Trieb*, life, and subjectivity, as well as the dynamic of contradiction and its *Aufhebung* which determines the relation between the finite and the infinite—it is this centerpiece of his theory that is utterly missed by the un-dialectical representations that appeal to unrelated determinations "such as *Trieb, Instinkt, Bedürfnis*, etc." (Enz. §359 Anm.). What the common presentations of organic nature miss despite their appealing to the same terminology is precisely the conceptual, living, hence, for Hegel, dialectic-speculative dimension of the natural phenomenon under investigation. Since such representations are unable to see that what those determinations entail is the dialectical force of contradiction in existence, they remain at the level of abstract and "empty formalisms" (see Schelling: Enz. §359 Zusatz, Hegel 1969–1971, TW 9, 472). Spinoza, on his part, while addressing the "absolute identity of life" is also unable to conceive its inner contradiction, hence to recognize that in animal life is contained the "*springenden Punkt der Selbstheit*, the principle of self-movement, the diremtion of oneself in oneself in general" (Enz. §359 Anm.). Spinoza's *conatus* is not dialectical (or not dialectical enough), hence is not truly *Trieb*. Hegel's position here is that what is new in his own account of the animal organism is not the terms employed but the method of which such terms are functions and by which they are connected (see, for instance, the *dialectical* connection *Mangel/Trieb* as negation of negation). At stake is the methodological difference that fundamentally changes the nature of the object that is being presented. It is the difference that separates a dead content from living, self-organizing life—and ultimately, the difference between substance and subjectivity.

While the dialectic *Bedürfnis/Instinkt* in which the tension *Mangel/Trieb* develops in the animal organism opens it up to the external environment, it does not lead it beyond such exteriority—it does not push the organism beyond its original determination and limit.[8] Herein lies the difference between nature and spirit. In the Psychology of subjective spirit, *Trieb* is introduced to characterize "practical spirit". This is spirit engaged in an activity that is properly the self-determination and self-production arising from the results of its theoretical activity. Human drive does not arise from instinct as in the animal but from theoretical intelligence. It is no longer *Naturtrieb* or *Bildungstrieb* but is the transformative self-production whereby spirit progressively appropriates its other by becoming its other—first of all its own natural exteriority and corporeality. Unlike the *Trieb* of the animal organism, spiritual or human *Trieb* is, directly, the "will"—*Wille*. But according to "the form of the content" it is the "still *natural* will, immediately identical with its determination—*Trieb und Neigung*" (Enz. §473). As drive, the will is still immersed in nature, in its inclinations and passions, hence is still a "dependent, unfree essence" (Nürnberg Enz. §178, Hegel 1969–1971, TW 4, 58). Quite in the sense presented in the logical method, *Trieb* is the *beginning* of spirit's voluntary action hence the beginning (and *only* the beginning) of its subjective freedom. As such, it indicates the abstract space of possible action aimed at satisfying inclinations, passions, interests, and feelings; but it also expresses the will's formal purposive activity (Enz. §475). In the sphere of "practical spirit", the drive is no longer instinctual activity

[8] See for this Bonito Oliva (2002).

determined by a need perceived as "lack". It is rather the activity determined by an "ought to"—a *Sollen*—(Enz. §473) that in and of itself already establishes the distance separating the human will from nature and opens up to spirit's action of shaping itself and its incipient freedom in a second nature. Human need, hence the drive to overcome it indicates a "lack" only to the extent that it is inadequate to what spirit assumes as a *Sollen*—as its own practical *Sollen*. This also implies that spirit's *Trieb*, despite its presenting itself as an immediate, is always already mediated (by the "ought to" that directs it but which the drive truly presupposes). Spirit's drive expresses the fact that the will "ought to posit the correspondence of its inner determination and its existence" (Enz. §473). The fulfilment of the drive is spirit's practical activity and is the beginning of spirit's realization in the objective world—the condition for the transition to the sphere of objective spirit. Indeed, while the "subject" is the very "activity of satisfying its drives", this satisfaction is the "translation out of the subjectivity proper to the content, which in this regard is the purpose, into objectivity". This objectivity is no longer the exteriority that limits the animal organism and to which the animal drive merely reacts. It is, instead, the locus of spirit's objective freedom. Embodying, yet again, the fundamental drive of the logical idea to find itself in everything through its own action, the *Trieb* of practical spirit is now translated into an objectivity "in which spirit reconnects itself with itself (*sich mit sich selbst zusammenschliesst*)" (Enz. §475). This is its freedom. *Trieb* is the force that guides spirit's *subjective* liberation from and within nature by bringing forth, as Hegel puts it, the *Befriedigung* (satisfaction) that establishes a (first) "*Frieden* (peace) between subject and object" (Enz. §204 Anm.). But it is also the beginning of spirit's *objective* freedom as its actualization in the world of its own social, political, and historical productions.

Consistently with my aim in this chapter, namely, to bring to light the systematic importance of Hegel's unique extension of the concept of drive to the Logic as the methodological condition for the development of the reality of *Trieb* in the philosophy of nature and spirit, these latter considerations touch only on the beginning of spirit's development of the drive that constitutes its freedom. This development unfolds through the subjective forms of "practical spirit" but continues further within spirit's objectivity in social and political institutions and world history. Throughout this itinerary, spirit's drive marks the points of contradictory tension but also mediation of nature and freedom, alien influence and self-determination, contingent individuality and the universal—the universal of natural life but also, successively, of morality and ethical life. Most pointedly, however, spirit's trajectory in this regard embodies the claim that Hegel establishes at the end of the Logic: the idea is not only "the highest power (*Kraft*) or properly the only and absolute power of reason, but *it is also the only and highest Trieb to find itself in everything*" (Hegel 1969–1971, TW 6, 551–552). For, ultimately, spirit's *Trieb* just as the animal *Trieb* has its root in—and is the actual embodiment of—the logical *Trieb* of the idea.

References

Aristoteles. 1831. Peri Poietikes. In *Aristoteles Graece*, ed. Immanuel Bekker, 1447–1462. Berolini: Reimer.

Bonito Oliva, Rossella. 2002. La traduction spirituelle du *Trieb* dans la philosophie hegélienne de l'esprit. *Revue Germanique Internationale* 18: 187–199. https://doi.org/10.4000/rgi.922.

Hegel, Georg Wilhelm Friedrich. 1969–1971. *Werke in 20 Bänden*. Frankfurt am Main: Suhrkamp. (TW).

Nuzzo, Angelica. 2012. *Memory, History, Justice in Hegel*. London: Palgrave Macmillan.

———. 2018a. *Wirklichkeit and Wirkung*: Hegel's Logic of Action. In *Wirklichkeit: Beiträge zu einem Schlüsselbegriff der hegelschen Philosophie*, ed. Luca Illetterati and Francesca Menegoni, 207–226. Frankfurt am Main: Klostermann.

———. 2018b. *Approaching Hegel's Logic, Obliquely: Melville, Molière, Beckett*. Albany: State University of New York Press.

15

Trieb and *Triebe* in Schopenhauer's Metaphysics of Nature

Marco Segala

1 Introduction

An arational tendency toward nothing but perpetual striving—this might be a good description of Schopenhauer's conception of the metaphysical will. Such a general view of the relation of the will to unstoppable struggling and desiring would admit of several distinct inquiries into ethics and metaphysics—and all of them would mention the notion of "drive" (*Trieb*). It is a term which often recurs in Schopenhauer's texts: it connects them with the philosophical tradition and the life sciences, it highlights their departure from the tradition itself, and it refers to psychological functions in a way that explains why Freud was indebted to Schopenhauer for his own interpretation of *Trieb*.[1] Schopenhauer was well aware of the conceptual import of "drive", and careful when using it. The aim of this

[1] See, for example, Schmidt 2004.

M. Segala (✉)
Dipartimento di Scienze Umane, Università di L'Aquila, L'Aquila, Italy
e-mail: marco.segala@univaq.it

M. Kisner, J. Noller (eds.), *The Concept of Drive in Classical German Philosophy*,
https://doi.org/10.1007/978-3-030-84160-7_15

chapter is to analyze how he employed and developed it in his system, and particularly in his metaphysics and philosophy of nature.

The theses I intend to maintain here are related to (a) the conceptual role that the *Trieb* acquired in Schopenhauer's thought and (b) how the notion of *Trieb* found its place in his metaphysics of nature. I'll lay out these main themes in four sections. The first one details Schopenhauer's view about the necessity of distinguishing the *Trieb* from the will. His account of the former as a manifestation of the latter in nature exhibits a character of the *Trieb* that should not be overlooked: it was meaningful for the philosophical understanding of nature, but not for the interpretation or comprehension of the will. Such a distinction is less clearly expressed than implicitly admitted in Schopenhauer's works, but it was explicitly enunciated in some passages of his correspondence that I am bringing to the attention of the reader. Moreover, such a distinction between *Trieb* and *Wille* and the relevance of the former for the comprehension of the interplay between natural phenomena suggest a further interpretation of its meaning in Schopenhauer's philosophy of nature. I contend that the discourse on the "drive" and the insistence on the plurality of "drives"—as opposed to the irreducible unity of the will—must be read as a further argument in favor of anti-reductionism—a view that Schopenhauer defended with vehemence but is often overlooked by commentators.

The second section proposes an answer to the historical question about the provenance of the notion of *Trieb* in Schopenhauer's thought. As the *Trieb* displays a character of the will, the former is conceived as related to the latter—and Schopenhauer also encouraged this interpretation. I propose a different story: Schopenhauer acknowledged the importance of the notion of *Trieb* at a time when he had not yet conceived of his metaphysics and not even his vocation for philosophy. It was at the time of his encounter with Blumenbach and the Bildungstrieb, as a student in Göttingen (1809–1811). I contend that the way Blumenbach discussed the Bildungstrieb guided Schopenhauer in elaborating his own notion of *Trieb*, which would later contribute to his philosophical interpretation of nature as objectivation of the metaphysical will.

The third and the fourth sections propose a reading of the application of the concept of *Trieb* to Schopenhauer's philosophy of biology. The third section reconstructs the way Schopenhauer confronted the challenge of teleology in animal instinct and *Kunsttrieb*, and shows that he arrived at a satisfying interpretation only in the second volume (1844) of *The World as Will and Representation*. The fourth section analyzes the most famous application of Schopenhauer's view of *Trieb*, namely the sexual drive in organic life and its effects in human society as "sexual love". These two sections display the interpretative power of Schopenhauer's conceptualization of *Trieb* as a fundamental and versatile notion of his philosophy of nature.

2 *Trieb*, Philosophy of Nature, and Anti-reductionism

In the second book of *The World as Will and Representation* nature—which comprehends human beings—is described as the "objectivation" of the will: "the appearance of that one and indivisible will that is the in-itself of all things and whose gradual objectivation is this whole visible world" (WI, 165/162). In themselves and according to their "innermost essence", stones, plants, and animals are nothing but "what we find immediately in ourselves as will" (WI, 126/130). In nature the will becomes object of cognition (representation) under the transcendental forms of space, time, and causality—an infinite variety of beings ordered according to "gradually increasing degrees of clarity and perfection" (WI, 199/191). Such a genesis of the world marks every single phenomenon—thus, the will "itself is present whole and undivided in every single thing in nature, in all of life" (WI, 153/154)—and is determined by the conflict of the will within itself (*Entzweiung mit sich selbst* or *Selbstentzweiung*, WI, 174–175/171–172). Struggling with itself, the will becomes other from itself and object of representation, and each stage of objectivation is a provisional appeasement of the ubiquitous struggle that dominates the natural world.

The world as will explains the intrinsic dynamicity and mutability of natural phenomena, organic life, and human existence. Everything inexorably changes without restraint or following a pre-ordained plan. The will appears as a *blinder Drang* (WI, 179/175), "as a purely blind impulse to exist" (WI, 186/181), and everywhere we observe the ever-occurring tendency of nature to avoid stasis. The word that Schopenhauer chooses and which better expresses the activity that the will brings into nature is *Trieb*. While the cognate *Drang* is generally related to describe the intrinsic activity of the will, *Trieb* catches the qualities, characters, and forces of the natural world as determined by the will: "perpetual striving" (WI, 368/338), blind tendency that appears "with the utmost certainty and infallibility" (WI, 179/175), will to life as "an inexhaustible driving-mechanism, an irrational drive" (WII, 409/373), and "the organic drives (*Getriebe*)" of the living bodies (N, 2/324; 9/331; 29/347). As for the human world, "[W]e will find here too that it is only a blind urge, a completely groundless, unmotivated drive" (WII, 407/372). Human passions squander "the majestic strength of great individuals" (WI, 216/206), "blind will comes forward as the drive to life, love of life, courage to face life" (WII, 410/374), language is determined by an "inner impulse (*Trieb*)" related to the process of thinking (N, 98/404) and even cognition "only ever comes into action under the impulse of willing" (WI, 221–222/211) and mostly appears "as a mere mechanism, a means for the preservation of the individual and the species as much as any organ of the body" (WI, 181/177).

It is important to underline that notwithstanding some concessions to rhetorical exigencies Schopenhauer resorted to a precise terminology: while *Trieb* describes the activity of the will in nature, *Drang* is mainly paired with the dynamics of the will in itself.[2] This is very clearly expressed by passages like the following: "without me as the subject of cognition, the thing cognized is not an object but rather mere will, blind urge (*blinder Drang*) [...]. As soon as cognition, the world of representation, is suppressed, absolutely nothing is left but mere will, blind urge (*blinder Drang*)" (WI, 212–233/203). Moreover, in a series of letters around the mid-1850s Schopenhauer was explicit about *Trieb* as a relevant notion to describe phenomena but not to be confounded with the will as the thing in itself. Referring to Immanuel Hermann Fichte's *Anthropologie* (1856)

[2] *Drang* is also used when the focus is on the effect of the will on human beings: *Drang der Wünsche* (WI, 231/220; 233/221), *Willensdrang* (WI, 231/220), *Drang des Wollens* (WI, 239/227).

and Karl Fortlage's *Genetische Geschichte der Philosophie seit Kant* (1852) and *System der Psychologie* (1855), Schopenhauer refused the equivalence between *Wille* and *Trieb* that these authors had established.[3] According to him, that equivalence attributed metaphysical qualities to a property of nature which, as expressed by the verb *treiben*, "indicated a *vis a tergo*" (GBr, 6.6.1856, 394).[4] He also made explicit that *Trieb* was "an unsuitable word, an image taken from the objective world" (GBr, 3.11.1855, 376) and accused those authors' use of *Trieb* as a covered plagiarism of his own notion of *Wille*: Fortlage "wants to appropriate my doctrine by putting *Trieb* instead of *Wille*" (GBr, 27.2.1856, 385).

As on other occasions, Schopenhauer's letters are helpful in clarifying notions that otherwise could be debatable or more open to interpretation.[5] The term *Trieb* was not a synonym of will and even less a figurative way to denote the will as a metaphysical entity. It was a notion that expressed the natural dynamics of a vast area of phenomena and was particularly useful in describing and defining organic activity. Schopenhauer mentioned a *Lebenstrieb* (WII, 410/374) and a "drive to self-preservation" (WII, 338–39/311), and assessed the importance of a Bildungstrieb in living organisms. When he observed animal and human life, he noted that "deeds and drives" (*Thun und Treiben*, WI, 186/181) and "pulls and drives" (*Zuge und Triebe*, WII, 641/575) express the intrinsic character of each species. Moreover, he further specified these drives in compound concepts relating to the apparently intelligent instinctual activities of some animals (*Kunsttrieb*) and the reproductive drive—*Befruchtungstrieb*, *Begattungstrieb*, *Zeugungstrieb*, and *Geschlechtstrieb*.

These different kinds of *Trieb* describe different fashions of analyzing the dynamics of the living beings. According to Schopenhauer's philosophy of nature, natural forces were the first level of the will's objectivation and, from the metaphysical point of view, they were interpreted as Platonic Ideas (WI, 154/155, 166/164–165). As such they were plural—a specification on which Schopenhauer insisted as a response to the then

[3] A brutal assessment of Fortlage's 1852 book is in GBr, 10.6.1852, 283, where its content is qualified as "falsch, schief und schlecht".

[4] GBr, 12.7.1852, 284 explains that an interpretation of the will as "substance" would be wrong.

[5] This is a main point of Invernizzi's view of Schopenhauer's correspondence in Invernizzi 1986, 254–258.

widespread reductionist assumption among scientists. He refused the possibility to explain one force through another (WI, 168/166), and in particular to reduce the forces responsible for the physiology of the living to the forces of the inorganic nature. He adamantly opposed this view—"physiological functions are never reducible to chemical or physical processes" (WI, 146/148)—and judged as stupid the attempt "to explain the phenomena of life using physical and chemical forces, which in turn are supposed to come from the mechanical operations of matter, the position, shape and movement of imagined atoms" (WI, 146/148). Such an aversion to mechanism became central in the treatment of matter and materialism in the second volume of *The World as Will and Representation*, where materialism was condemned also because it denied the plurality of forces "by seemingly and supposedly reducing them all, and ultimately even the life force, to the merely mechanical efficacy of matter" (WII, 357/326).

The *Trieb* was introduced as a conceptualization which described both the conflict as it appears in nature at each stage of the will's objectivation and the striving toward the overcoming of such a conflict, which would bring about a subsequent stage (WI, 179/175). Even the communication of motion was described as "an expression of the basic striving of the will in all of its appearances" (WII, 338–339/311) and gravity as the force driving the bodies toward the surface of the earth (N, 80–81/389).[6] But Schopenhauer did not indulge in using this concept in the domain of the inorganic world; instead, he restricted its meaning by mainly referring to the variety of physiological processes related to life, instinct, and reproduction: "everything strains and drives (*drängt und treibt*) towards *existence*, toward *organic* existence if possible, i.e. towards *life*" (WII, 399/365). I suggest that it was a choice apt to emphasize the anti-reductionist stance of his philosophy of nature. The notion of *Trieb* was preferably employed to express in an intelligible way the general character of the action of the will in the living forms, even without the precise identification of the forces responsible for the different physiological operations of the organisms. Moreover, the mention of different kinds of

---

[6] Actually, this argument came from the quotation of Herschel's *Treatise on Astronomy* that introduced the chapter "Physische Astronomie" in *On Will in Nature*.

Trieb added more weight to Schopenhauer's view about the plurality of forces and his anti-reductionist arguments.

3 From *Bildungstrieb* to *Trieb*

The way Schopenhauer distinguished *Trieb* and *Wille* is explained not only by the theoretical reasons presented in the previous section but also by the story of Schopenhauer's first encounter with the notion of *Trieb*. Such a story merits a detour and requires the recapitulation of an important chapter in the history of the life sciences—the introduction of the notion of *Bildungstrieb* by Johann Friedrich Blumenbach. In 1781 he published a booklet (less than 90 pages) entitled *Über den Bildungstrieb und das Zeugungsgeschäfte* (On the Formative Drive and the Operations of Reproduction). After decades of heated debates between mechanism and vitalism concerning the notion of life, and of preformation and epigenesis as theories of generation, Blumenbach contributed with a major breakthrough and introduced the view of a "life-long active drive (*Trieb*)" which "bestows on creatures their form, then preserves it, and, if they become injured, where possible restores their form. Such a drive (or tendency or effort, however one wishes to call it) is completely different from the common features of the body generally and also from the other special forces (*Kräften*) of organized bodies in particular. It appears as one of the first causes of all generation, nutrition, and reproduction, and in order to avoid all misunderstanding and to distinguish it from all the other natural forces, I refer to it with the name of *Bildungstrieb* (Nisus formativus)" (Blumenbach 1781a, 12–13).

At the time of this first formulation, the notion of *Bildungstrieb* embodied the primordial character of life as a tendency toward an ordered structure. It was not a force, but it grounded the forces—like sensibility, irritability, reproduction—that generally explained the operations of organisms. It had a "comprehensive architectonic character" (Richards 2000, 19) and manifested itself through those operations in which the living forms expressed their organizing tendencies. Its ability to direct the process of giving an ordered form to an organism, orchestrate the different physiological functions, and harmonize the operations of the different organs made it more versatile than the more specialized forces invoked by previous naturalists and philosophers.[7] Working out the *Trieb* responsible for such a versatility, Blumenbach was able to extend the role of the formative drive in explaining the variability of forms in nature. He gave it a major function in anthropology—thus explicating the varieties of races and individuals in humans—and natural history, where the subject was the varieties of species and the creation of new species after the deluge. Moreover, Blumenbach applied the formative drive to the investigation of similarities among living forms while promoting the rise of comparative anatomy.[8]

In the following years Blumenbach revised the concept and in the second edition of *Über den Bildungstrieb* (1789) he proposed a new interpretation of its nature: he made clear that the formative drive was a force that from a philosophical point of view must be considered analogous to Newton's gravity. Like gravity, which is mysterious about its cause but has great explanatory power, *Bildungstrieb* "should serve, no more and no less, to signify a force whose constant effect is recognized from experience and whose cause [...] is for us a *qualitas occulta* [...] It holds for all these forces what Ovid says:—*caussa [sic] latet, vis est notissima*" (Blumenbach 1789, 25–26).

In the end, the *Bildungstrieb* was a force—but not the cause—that explained the distance of the organic from the inorganic realm, and yet its intrinsic nature as tendency qualified its difference from the traditional forces invoked to explicate the peculiar nature of the living forms. The *Bildungstrieb* was less like a traditional force and more like a complex of factors which act continuously and concur to give form and stability to an organism—it provided the explanation of the goal-directed organization of living beings.[9]

---

[7] On the main views in the eighteenth century, see Monti 1990 and Zammito 2017, especially chapters 3–5.

[8] The presence of the *Bildungstrieb* in anthropology is evident in the second edition of *De generis humani varietate nativa* (Blumenbach 1781b, 1–2). For the *Bildungstrieb* in natural history, see *Beyträge zur Naturgeschichte* (Blumenbach 1790, 25). Its application to comparative anatomy appears in *Handbuch der vergleichenden Anatomie* (Blumenbach 1805, 65, 471).

[9] On this aspect, see Gambarotto 2018, 10–14. Among historians there has been a lot of discussion about the meaning of the *Bildungstrieb* for Kant's analysis of teleology: see, for example, Zammito 2012—but examining this subject would be beyond the scope of the present contribution.

The circulation of the theoretical notion of *Bildungstrieb* in German philosophy and biological sciences at the turn of the nineteenth century was impressive. The novel concept was mentioned even by Kant in the *Critique of the Power of Judgment* (section 81), and rapidly spread. Among the physicians and physiologists who discussed or incorporated it in their discussions on the life force(s), we can recall Christoph Wilhelm Hufeland, Joachim Dietrich Brandis, Carl Friedrich Kielmeyer, Theodor Georg August Roose, and Johann Christian Reil. Among the philosophers, the notion of *Bildungstrieb* was disseminated by the commentators of Kant's third *Critique* and discussed by Johann Gottfried Herder, Salomon Maimon, Carl Christian Erhard Schmid, Christoph Gottfried Bardili, Fichte, Schelling, Hegel, and Fries. The explorer and anthropologist Georg Forster—and also Alexander von Humboldt and Johann Wolfgang Goethe—approached the *Bildungstrieb* as a regulative notion which connected anthropology to the creative (artistic and cultural) drive in humanity. Romantic thinkers like Schiller, Hölderlin, Novalis, Friedrich Schlegel, Jean Paul, and Joseph Görres exported Blumenbach's biological notion into the realm of aesthetics as an explanation of the artistic creativity.[10]

Schopenhauer was among the many authors who analyzed and discussed the *Bildungstrieb*. Already in the first edition of *The World as Will and Representation* he mentioned it as explanatory of the apparent intentions manifested by organic nature—like the snail's house, which is not certainly the effect of a cognitive intention but of the will operating "blindly [...], as a formative drive directed outwards" (WI, 137/140). It is evident, however, that his was not a cogent formulation—certainly not comparable to the more refined conceptual elaborations provided by Goethe or Schelling.[11] Schopenhauer remained cautious about giving weight to the *Bildungstrieb* in his philosophy, but that does not mean that it had a marginal role in his work. On the contrary, I want to emphasize that it had a very important meaning—less related to the content of that

[10] A comprehensive overview of the *Bildungstrieb* during the Goethezeit is provided by Fabbri Bertoletti 1990.

[11] See Zammito 2017, chapter 10–11 and Cooper's contribution in this volume.

notion, however, and more to the methodological lesson that Blumenbach had taught through it.

Schopenhauer learnt that lesson—not metaphorically but literally—as Blumenbach's student between 1809 and 1811. He had enrolled in the faculty of medicine at the university of Göttingen and he did not miss a single one of the four courses delivered by the great scientist during those two years.[12] As a young man (he was 21), he had the chance to attend the lectures of one of the *praeceptores Germaniae*, renowned for his formidable accomplishments and his uncontested mastery in anatomy, physiology, anthropology, and natural history.[13] I insist on this celebratory language, because it helps us imagine Schopenhauer's awe for the professor. It was an important relation with important consequences that lasted many years—Schopenhauer even asked for Blumenbach's advice when it was time to apply for a university position after the publication of *The World as Will and Representation.*[14]

The peculiarity of Schopenhauer's encounter with Blumenbach is worthy of attention and explains the particular character of the former's interest in and use of the notion of *Bildungstrieb*. As mentioned before, it was mostly of methodological interest, but nonetheless it was relevant and meaningful for understanding what is pertinent here—the value and role of the notion of *Trieb* in Schopenhauer.

Learning about the *Bildungstrieb* from the voice of his creator had two important consequences. The first one regards Schopenhauer's view of the relationship between science and metaphysics, and in particular the notions of ground, cause, force, and phenomena. As sketched earlier, Blumenbach's view of the *Bildungstrieb* had evolved from being a primal, original tendency of any organic matter, which manifested itself through forces, into a secondary cause—a force, produced by a mysterious cause, that could be recognized by its phenomenal effects. Schopenhauer would

[12] Natural history (winter semester 1809–1810), Mineralogy (winter semester 1809–1810), Comparative anatomy and physiology (winter semester 1810–1811), and physiology (summer semester 1811). The notion of *Bildungstrieb* appears in the lectures on physiology: see Stollberg and Böker 2013, 126 and 134.

[13] On this definition of the Göttingen professors as "masters of Germany", see Marino 1995.

[14] Their correspondence is published in *Werke* (Deussen), 265–267 and 275–276. On the meaning of Blumenbach's teaching for Schopenhauer's philosophizing, see Segala 2013, section 5–6.

later recall such a distinction while discussing the *Bildungstrieb* in *On Will in Nature*, when he wrote that "natural forces, vital forces, and formative drive" are names that hide a reality which is unknown to and unknowable by the sciences (N, 4/326 and 25/344). It was a view already discussed in *The World as Will and Representation*, where several passages made clear the distinction between the scientific notion of force and the philosophical concepts of ground and cause.[15] Like Blumenbach, Schopenhauer described the force as *qualitas occulta* (WI, 97/107) and, more specifically, as the product of the "intuitive cognition of the objective world, [...]. It is abstracted from the realm governed by cause and effect, and thus from intuitive representation, and signifies simply the causal nature of a cause at the point where, aetiologically, it can do no more explanatory work but rather is the necessary presupposition of all aetiological explanation" (WI, 133/136–137). For this reason he claimed that a force requires a further explanation—not a cause, however, as stated by Blumenbach, but an ultimate ground, "the unfathomable (the groundless, i.e. the will)" (WI, 144/146).[16] He followed Blumenbach methodologically: he considered the necessity of something beyond the force and found it in the "groundlessness of the will" (WI, 135/138), which provided the ultimate explanation of every force.

The second consequence of Blumenbach's teaching was even more important: it made available to the young Schopenhauer the notion of *Trieb* and its flexibility—Blumenbach mentioned the *Geschlechtstrieb*, too[17]—and gave him the occasion to assess this conceptual tool and its use in the domains of science and philosophy of science. As briefly summarized earlier, the *Bildungstrieb* introduced an innovative view of organic forces as independent from, but related to, mechanical forces that had a deep impact in defining and promoting the development of the life

[15] See WI, §§ 15, 17, 22, and 24.

[16] The German text revolves around the term *Grund*: "das Unergründliche (Grundlose, d. i. Wille)". About the notion of cause, Schopenhauer remained on Kant's position and recalled that a cause is merely a relation between phenomena and is therefore inapplicable to forces, which are beyond the realm of phenomena. See WI, 16/35 about the inconceivability of a cause beyond the world of representation.

[17] The *Geschlechtstrieb* was mentioned in Blumenbach's lectures on physiology and on comparative anatomy: see Stollberg 2013, 92 and 104.

sciences and the philosophy of biology in Germany around 1800.[18] At the time of his first scientific education in Göttingen, Schopenhauer was not aware of such an important role yet, was still uncultivated in philosophy or philosophy of science, and was still very far from the metaphysics of the world as will. When he learned about the *Bildungstrieb* and the *Geschlechtstrieb* from Blumenbach himself, he had the privilege of immediately appreciating the distinction between *Trieb* and *Kraft* and the versatility of the former.[19] Even if Blumenbach had eventually viewed the *Bildungstrieb* as a force, it was reasonable to infer that Blumenbach's discourse about the occult quality behind it was related to its nature as *Trieb*—the common quality shared with the sexual drive, which, however, was not presented as a force. This was the path followed by Schopenhauer when he would later interpret the operations of the will in nature as expressed by a *Trieb*. Even if in the end Schopenhauer did not adopt the notion of *Bildungstrieb* to a great extent—as other philosophers of the *Goethezeit* did—and merely listed it as one of the several natural forces which remained mysterious and necessitated a philosophical explanation, he nonetheless applied Blumenbach's lesson and looked at the *Trieb* as something more general and profound. Actually, as presented in the previous section, it became the most appropriate notion to express the intrinsic dynamics of nature as objectivation of the will.

4 Kunsttrieb

The reason why Schopenhauer showed little enthusiasm for the Bildungstrieb poses an interesting question. After all, it was the notion developed by his teacher in Göttingen, included by Kant in his third *Critique*, and expanded by Goethe, who also was a mentor of Schopenhauer's in Weimar between 1813 and 1814. Considering that three major figures of his education had supported and enriched the Bildungstrieb, Schopenhauer had many good, intellectual reasons to

[18] See Gambarotto 2018, chapter 1.

[19] Zammito 2017, 212 underlines the importance of the distinction between *Trieb* and *Kraft* not only for Blumenbach but also for Kant, Schelling, and Goethe.

embrace the notion—but he did not. I suspect that on this he followed Schelling: he was not sympathetic about a philosophical use of the notion that would serve as an explication of the relationship between organic and inorganic matter. He shared Schelling's view that the *Trieb* was something immanent to nature and its intrinsic activity and he resisted the idea to consider it—or any manifestation of it—as a philosophical explanation. He was convinced that a good scientific notion—like the Bildungstrieb—would not also be a good philosophical foundation of nature. It could illuminate some philosophical questions, but it did not provide a philosophical answer.

This view seems to be confirmed by his wider and deeper discussion of another version of the notion of *Trieb*—the *Kunsttrieb*. In comparison with the Bildungstrieb, it was a concept with a longer and not less important tradition in the domain of the life sciences—and it also was a good candidate to a philosophical interpretation. It was defined and inquired by Hermann Samuel Reimarus in *Über die Triebe der Thiere* (1760), a book Schopenhauer read and whose second edition (1762) he acquired for his library. Against mechanism, materialism, and impiety, Reimarus proposed an interpretation of animal instincts (*Triebe*) as tendencies which pursued without cognition a beneficial goal for an individual or its species, thus exhibiting the providential intervention of God in nature. He called *Kunsttrieb* this particular kind of instinct, where the word *Kunst* referred to the artifice and the skillfulness necessary to achieve an objective.[20]

Schopenhauer followed Reimarus's views. As *Kunsttriebe* are goal-oriented and benefit organisms and their species, he considered generation (*Zeugung*) and sexual desire in humans as kinds of *Kunsttrieb* (WII, 587/528–529 and 619/557). Like Reimarus, who had supported his arguments with several examples of the purposiveness of the instinctual activity, Schopenhauer was fascinated by animal behaviors—"the spider's web, the bee's honeycomb, the termite's hill, etc." (N, 39/355)—which seemed to pursue beneficial goals without the intervention of knowledge. He noticed that *Kunsttriebe* "are not guided by motives or cognition, but seem to perform their tasks from abstract, rational motives" (WI,

[20] See Zammito 2017, 138–44 and the contributions 2 and 4 in the present collection.

180/176) and concluded that "the instincts and creative drives of animals show us at once that the will is at work even where it is not guided by cognition" (WI, 136/139). Moreover, he adopted Reimarus's position that viewed *Kunsttriebe* as operations that worked even better than those promoted or planned by the knowledge faculties when pursuing an objective. It was a powerful argument against the traditional superiority of intellect and reason—on the contrary, the intellect was "secondary" and knowledge was subjected to "illusion and deception", whereas instinct and *Kunsttrieb* "in the absence of cognition" exhibits "infallibility" (WI, 180–181/176) and act "infinitely better and more perfectly than things that take place with the assistance of intellect" (WII, 304/282).

Schopenhauer's interest in the *Kunsttrieb* and his enthusiasm in promoting its importance in the system are quite intriguing. It is certainly related to the role he gave to the notion of *Trieb*, but there was more than that. We have also to consider that in Reimarus the *Kunsttrieb* was an original inclination posited by God. It was designed to interact with the natural course of life and to introduce harmony and order in the absence of knowledge and intention by the agents. It was an expression of the physico-theology that Schopenhauer despised and debated in the chapter "Comparative Anatomy" of *On Will in Nature*. He was well aware that "the obvious suitability of any animal to its way of life, [...] the purposiveness without exception, the evident intentionality in all parts of animal organism" (N, 37/353) had traditionally offered the best argument in favor of a theological explanation of nature. Yet he had an alternative, atheist explanation—the will and the *Kunsttrieb* still appeared as its best evidence: "the physico-theological proof has already been rendered powerless by the empirical observation that the works of the *Kunsttrieb* in animals [...] are all constituted as if they were the result of a purposeful concept, as if they originated in far-reaching foresight and rational deliberation; whereas they obviously are the work of a blind drive, i.e. of a will not guided by cognition" (N, 39/355).

It was a bold move, even if less convincing than he would admit. Schopenhauer took the *Kunsttrieb* and put it into a new conceptual framework—not an evidence of the existence of God, as it required a divine intervention in nature, but the proof of a will that manifests itself in the preservation and promotion of individuals and species. The many pages of the chapter on "Comparative Anatomy"—which are filled with examples of "the quite amazing purposiveness and harmony in the animal's inner mechanism"—take the reader to the conclusion that "the animal is just *its will itself*" (N, 54/367), and the intermediate passage of the argument was provided by the *Kunsttrieb*, which made "all of these arrangements comprehensible to ourselves by way of what is apparent in" it (N, 47/362). The will as a metaphysical entity explained the intrinsic operations of nature and those actions of "quasi-cognition [...] not motivated by the brain" like "instinct and the *Kunsttrieb*" (WII, 291/270). But as the will was without cognition and intentions, how could it manifest activities apparently determined by knowledge and intentions?

Notwithstanding its firm determination, however, the argument was still lacking a good explanation of how the intentionality of the *Kunsttrieb* could follow from a metaphysical entity utterly indifferent to rationality and planning. Such an explanation was the subject of chapter 27 ("On Instinct and Creative Drive") in the second volume of *The World as Will and Representation.*

The *Kunsttrieb* had such an important role that it deserved a special treatment. Schopenhauer made it clear at the beginning of the chapter why it was a unique feature: "with the creative drive of animals, it is as if nature wanted to hand the researcher on a plate an explanatory commentary on its operation according to final causes and on the amazing purposiveness of its organic productions that results" (WII, 390/357). He was addressing again the question of teleology, after the main analyses in section 28 of the first volume of *The World as Will and Representation* and chapter 26 of the second. Making consistent "the completely blind works of nature [...] with appearance of intentionality" (WII, 382/349) was a key element in his philosophy—where the irrational will was the ultimate explanation. Schopenhauer was aware that his intention of breaking the traditional bond between teleology and "speculative theology" (WII, 385/352) was revolutionary and blamed philosophers who after recognizing purposiveness in nature "break out at once into childish cries of 'design! design!'" (WII, 386/352–53). But the *Kunsttrieb* was a tough subject to tame within a non-theological system grounded on the irrational will.

In this text Schopenhauer engaged the challenge of offering a philosophical explanation of the *Kunsttrieb* which would be supported by scientific evidence. Instead of asserting that animals exhibited intentional activities without conceiving a goal—because of the aimless will—he reformulated the question admitting that a pinch of knowledge was present in any *Kunsttrieb*. The premise was that instinct—"a distinct drive of the will"—requires an "external circumstance" to be expressed; from this it followed that "the instinct is what is primarily active in the operation of the creative drive, although the intellect of these animals is active too in a subordinate way: instinct gives the universal rule; intellect gives the particular, the application, by presiding over the details of how the animal carries out its work" (WII, 391/358). As such a sprinkle of intellect was everywhere "in animals in the lower classes, particularly insects" (WII, 382/358), it became clear why *Kunsttrieb* was so effective and amazing among animals "with an extremely limited cognitive sphere" (WII, 382/358). Moreover, Schopenhauer proposed a corollary to his explanation by giving a reason for our surprise and inclination to invoke a divine intervention before the *Kunsttrieb*. It seemed to require a sophisticated knowledge that was clearly absent from lower animals, but in fact their tiny intellect was up to the task "because, guided by instinct, it has only to make good any gaps instinct has left" (WII, 396/361).

5 Sexual Drive

Like *Kunsttrieb*, another form of drive—this time in humans—exemplified the versatility of Schopenhauer's notion of *Trieb* as a primary expression of the manifestation of the will in nature: the sex drive. According to Schopenhauer, the major evidence of the incessant dynamics introduced by the will in the living nature as a constant restlessness was provided by the tendency to reproduction and sexual desire in the more complex organisms. The notion of *Trieb* became the fulcrum of Schopenhauer's analysis of this essential part of life, as it alone guarantees the perpetuation of life. The "enormous overabundance of seed and the great power of the fertilizing drive (*Befruchtungstrieb*)" were recalled as biological evidence of the unstoppable power that propels living beings to reproduction (WI, 325/302). The procreative drive (*Begattungstrieb*) was seen as powerful as hunger (WII, 403/368), and was exemplified by an experiment showing how "a fly poisoned with arsenic carries on procreating, from an unexplained drive, and dies in the act of procreation" (WII, 382/349). From the fact of "the urgent vehemence of the sex drive (*Geschlechtstrieb*) and its zealous adaptation to all circumstances and opportunities" (WII, 401/366) it appeared that the sex drive was "the focal point of the will" (WII, 268/250; 588/530), "the kernel of the will to life" (WII, 588/530), and "the most vehement of the desires, the wish of the wishes, our willing as a whole in concentrated form" (WII, 589/530).

Schopenhauer viewed the dynamics of the living nature as grounded on the species, which corresponded to one "determinate and fixed *level of the will's objectification*" (WI, 154/155) and were properly real, non-transient things: "the life of the individual is fundamentally only borrowed from the species and [...] all life force is, as it were, species force (*Gattungskraft*)" (WII, 585/527). Nature showed a deadly indifference for individuals and was directed "only to the continuation of the species and the general conditions of life, not to the continuation of individuals" (WI, 192/186). The preservation of the species was nature's ultimate "intention (*Absicht*)" (WII, 401/366) and procreation was therefore the "highest point" of any individual, whose life "fades slowly or quickly away" after having attained this goal (WI, 195/188).

The centrality of the sexual drive in animals was examined in the chapter "Life of the Species" of the *Supplements* to *The World as Will and Representation*. Notwithstanding the importance of the digestive apparatus, the root of the individual's life was in the genitals, "because they connect the individual to the species in which it is rooted. [...] The sex drive should be viewed as the inner impulse of the [species] from which the life of the individual sprouts [...]: this is why the sex drive is so strong and why it comes from the depths of our nature" (WII, 584/527). Schopenhauer explained that among the physiological functions and organs, which serve the individual, only the vehement sexual drive—"which is the concentration of the entire animal essence" (WII, 586/528)—was evidence of the primacy of the species over the individual.

Schopenhauer's insistence on the primacy of the species was not only required by metaphysical reasons—the ontological superiority of the Ideas—but also by the necessityS of explaining the power of the sex drive in humans. If it were a genuinely individual impulse, a human could easily fight it through reason and dominate it. But as a necessity of the species, the sex drive would be impossible to keep at bay, as it would incarnate the real nature of each individual as a member of the species. This reading is supported by these astonishing definitions: "the human being is the sex drive made concrete" (WII, 588/530); "sexual desire [...] is the wish that itself constitutes the essence of the human being" (WII, 587/529); "sexual relations in the human world [...] are the invisible focal point of all our doings and dealings" (WII, 588/529).

Nonetheless the case of the human species required a more refined analysis. Schopenhauer described humans as the only animals where "each individual needs to be studied and fathomed on its own [...]. This phenomenon of the particularization of individual character distinguishes people from animals; and this is illustrated by the fact that in animals the sex drive (*Geschlechtstrieb*) seeks satisfaction without noticeable selection, while with people this selection is instinctively and unreflectively pursued to the point of violent passion" (WI, 156/156). The genitals of the humans were "objectified sex drive" (WI, 129/133), like it was the case for all of the other animals, but the sexual impulse was intertwined with many other emotions related to the psychology of the individual, and this is the reason why Schopenhauer gave it a separate treatment in the famous "Metaphysics of Sexual Love" (WII, chapter 44). Notwithstanding its inescapable tendency to the preservation of the species, human sexuality was also profoundly sentimental—and this is the reason why Schopenhauer chose the term "sexual love".

The importance of Schopenhauer's treatment of this theme for the contemporary biological and cultural views about sexuality cannot be underestimated. Roger Scruton observed that Schopenhauer's analysis of the relationship between individuality of the sexual desire and the "long-term impersonal strategy of the species" was characterized by "a directness and psychological penetration which no sociobiologist has been able to match" (Scruton 2015, 191). Schopenhauer explored several aspects of human sexuality and did not leave out delicate issues, like homosexuality, pederasty, and incest—which was explained as extremely rare, "an unnatural perversity of drives (*Perversität der Triebe*)" (WII, 606/545). According to him, love was nature's stratagem for pushing humans toward procreation. It appeared as "objective admiration [...] tinged with sublimity", but essentially it was not a sentiment of reciprocity, rather "possession, i.e. physical pleasure" (WII, 612/551). Rape and forced marriages were mentioned as the typical examples of the equation between love, possession, and satisfaction of the desired pleasure. "Being in love, however ethereal" was "rooted solely in sex drive" (WII, 610/549), and the "attraction of two lovers is in fact already the life-will of the new individual who they can and want to conceive" (WII, 613/552). The power of nature was so "much more powerful than that of the individual that the lover closes his eyes to all the qualities that disgust him [...]. This is why the ancients portrayed Eros as blind" (WII, 637/571).

The attention to "the unconscious considerations" (WII, 623/560) determining the reciprocal attraction between two partners was the subject of a thorough inquiry into the importance of their physical and psychological qualities. These were responsible of the "significant passion which acquires the air of something more noble and more sublime", which had the task to generate the illusion of an individual connection and need—though unconscious instinct and sex drive were directed to satisfy the real need of the species, namely the act of procreation.[21] It explained the variety of connections between will to life and *Geschlechtstrieb* on the one hand and love and pleasure on the other hand. It was finalized to the survival of the species but was subtly designed to satisfy individuals within an "absolutely distinctive and individual passion between two lovers" (WII, 614/553).

This focus on the individuality of each love relationship expressed the tone of Schopenhauer's inquiry into love and sexuality. Love and passion were defined by the qualities and nature of two individuals, and "the more perfect the mutual suitability of two people for each other [...], the stronger their mutual passion will turn out to be" (WII, 615/554). It

[21] Schopenhauer was aware that homosexuality and pederasty represented a serious counterexample to his view and in an Appendix added in the 1859 edition (WII, 643–651/576–582) he provided a more complex (and not homophobic) solution, instead of simply referring to a degenerated "tendency" (WII, 620/558).

seemed that nature was selecting "a particular and determinate individual that can only be conceived by this father with this mother" (WII, 630/566). Schopenhauer insisted on the individual character of love and sexuality, and meticulously described the different angles and nuances of falling in love, being in love, and the end of love. He finally admitted that the naturalistic explanations he was offering "will upset those ensnared by passion, yet if rational considerations can do anything against this, the principles I have discovered will be able to overpower the passions more than any other ones" (WII, 639/573).

The modern reader will probably object about Schopenhauer's treatment of the more sensitive themes connected to sexuality. His misogyny is evident,[22] and it could serve as counterarguments to Scruton's praise—even if when contextualized they might hurt less our feelings. But if we focus less on the content and more on the method, we can appreciate Schopenhauer's idea of establishing a naturalistic view of sexuality, which emphasized the main role of the *Geschlechtstrieb* and its service to the species. Against any Romantic idealization, Schopenhauer viewed love as mainly sexual, a biological need as essential as nutrition which determined actions and choices of individuals. As a consequence, morality and pure feelings appeared as effects of blind drives, and not as the original sources of affection and partnership.

6 Conclusion

As the *Trieb* exhibited the specific character of the will—its arational tendency to willing—and manifested itself in different ways in the phenomenal world, it became a multifaceted notion in Schopenhauer's philosophy—and therefore it offers to the interpreter various perspectives on it. In this chapter, I have investigated the role of the *Trieb* in metaphysics of nature, but it also might help to better understand moral agency—related either to individuals in everyday life or humanity in history.

[22] As an example, consider this passage: "the marital fidelity is artificial in man and natural in women, and female infidelity is much less forgivable in a woman […] because of its unnaturalness" (WII, 621/558).

With regard to metaphysics of nature, Schopenhauer described the *Trieb* as interacting with natural forces and contributing to activity and conduct of the living organisms. It was this specific organic function—together with its versatility in different domains of the life sciences and human biology—that exhibited the distinction between *Trieb* and will. Moreover, as I have suggested earlier, such a distinction was instrumental in both Schopenhauer's philosophy of nature and his views against reductionism. It is true that Schopenhauer did not specify a genetic process of the different kinds of *Trieb*—as he did for the forces, in the second book of *The World as Will and Representation*—and maybe this lack of specification offered some reasons to philosophers in the 1840s for speculating about a synonymy between will and *Trieb*. But he was adamant in opposing that interpretation, and we have now a historical reason for understanding that reaction: the reconstruction proposed here—about his notion of *Trieb* as derived from Blumenbach's Bildungstrieb—shows that it originally was a scientific subject not related to a metaphysics of will.

Last but not least, the notion of *Trieb* enriched Schopenhauer's conceptualization about philosophy of science, as it added a new element to the distinction between ground and causality in understanding the operations of nature. Like any force in organic nature, the *Trieb* directed animal activity but, differently from any force, it exhibited different manifestations when interacting with intellect (in both human and nonhuman animals) and reason (in humans). It was precisely this interaction with the higher faculties that gave the *Trieb* its peculiar status and would oppose the domination of drives in individuals. The chapter on sexual drive and human love in the *Supplements* exemplified the complexity of an ethical inquiry which would consider the interaction between biology and moral responsibility.

References

References to Schopenhauer's works are abbreviated as follows. Page numbers refer to the German edition and are followed by reference to the page number of the English translation as listed here.

GBr Schopenhauer, Arthur. 1978. *Gesammelte Briefe*. Edited by Arthur Hübscher. Bonn: Bouvier.

WI Schopenhauer, Arthur. 1972. *Die Welt als Wille und Vorstellung. Vier Bücher nebst einem Anhange, der die Kritik der Kantischen Philosophie enthält* (1859). In *Sämtliche Werke*, vol. 2, Arthur Schopenhauer, edited by Arthur Hübscher. Wiesbaden: Brockhaus.

Schopenhauer, Arthur. 2010. *The World as Will and Representation*, vol. 1. Edited by Christopher Janaway. Translated by Judith Norman, and Alistair Welchman. Cambridge: Cambridge University Press.

WII Schopenhauer, Arthur. 1972. *Die Welt als Wille und Vorstellung, welcher die Ergänzungen zu den vier Büchern des ersten Bandes enthält* (1844). In *Sämtliche Werke*, vol. 3, Arthur Schopenhauer, edited by Arthur Hübscher. Wiesbaden: Brockhaus.

Schopenhauer, Arthur. 2018. *The World as Will and Representation*, vol. 2. Edited and translated by Judith Norman, Alistair Welchman, and Christopher Janaway. Cambridge: Cambridge University Press.

N Schopenhauer, Arthur. 1972. *Ueber den Willen in der Natur* (1854). In *Sämtliche Werke*, vol. 4, Arthur Schopenhauer, edited by Arthur Hübscher. Wiesbaden: Brockhaus.

Schopenhauer, Arthur. 2012. *On Will in Nature*. In *On the Fourfold Root of the Principle of Sufficient Reason and Other Writings*, edited and translated by David E. Cartwright, Edward E. Erdmann, and Christopher Janaway, 303–475. Cambridge: Cambridge University Press.

Werke (Deussen) Schopenhauer, Arthur. 1929. *Der Briefwechsel Arthur Schopenhauers*, vol. 1. Edited by Carl Gebhart. München: Piper.

Blumenbach, Johann Friedrich. 1781a. *Über den Bildungstrieb und das Zeugungsgeschäfte*. Göttingen: Dieterich.
———. 1781b. *De generis humani varietate nativa*. Göttingen: Vandenhoeck.
———. 1789. *Über den Bildungstrieb*. Göttingen: Dieterich.
———. 1790. *Beyträge zur Naturgeschichte*. Göttingen: Dieterich.
———. 1805. *Handbuch der vergleichenden Anatomie*. Göttingen: Dieterich.
Fabbri Bertoletti, Stefano. 1990. *Impulso formazione e organismo: Per una storia del concetto di "Bildungstrieb" nella cultura tedesca*. Firenze: Olschki.
Gambarotto, Andrea. 2018. *Vital Forces, Teleology and Organization: Philosophy of Nature and the Rise of Biology in Germany*. Cham: Springer.
Invernizzi, Giuseppe. 1986. Schopenhauer attraverso il suo epistolario. *Rivista di Storia della Filosofia* 41 (2): 245–264.
Marino, Luigi. 1995. *Praeceptores Germaniae: Göttingen 1770–1820*. Göttingen: Vandenhoeck & Ruprecht.
Monti, Maria Teresa. 1990. *Congettura ed esperienza nella fisiologia di Haller: La riforma dell' "anatomia animata" e il sistema della generazione*. Firenze: Olschki.
Reimarus, Hermann Samuel. 1762. *Allgemeine Betrachtungen über die Triebe der Thiere: hauptsächlich über ihre Kunsttriebe: Zur Erkenntniss des Zusammenhanges der Welt, des Schöpfers und unser selbst*. Zweyte Ausgabe. Hamburg: Bohn.
Richards, Robert J. 2000. Kant and Blumenbach on the *Bildungstrieb*: A Historical Misunderstanding. *Studies in History and Philosophy of Biological and Biomedical Sciences* 31 (1): 11–32.
Schmidt, Alfred. 2004. Von der Willensmetaphysik zur Metapsychologie: Schopenhauer und Freud. In *Tugend und Weltlauf: Vorträge und Aufsätze über die Philosophie Schopenhauers (1960–2003)*, ed. Alfred Schmidt, 317–427. Frankfurt am Main: Peter Lang.
Scruton, Roger. 2015. *Sexual Desire: A Philosophical Investigation*. London: Bloomsbury Publishing.
Segala, Marco. 2013. Auf den Schultern eines Riesen. Arthur Schopenhauer als Student Johann Friedrich Blumenbachs. In *"...die Kunst zu sehn": Arthur Schopenhauers Mitschriften der Vorlesungen Johann Friedrich Blumenbachs (1809–1811), mit einer Einführung von Marco Segala, Arthur Schopenhauer*, ed. Jochen Stollberg and Wolfgang Böker, 13–40. Göttingen: Universitätsverlag Göttingen.

Stollberg, Jochen, and Wolfgang Böker, eds. 2013. *"…die Kunst zu sehn". Arthur Schopenhauers Mitschriften der Vorlesungen Johann Friedrich Blumenbachs (1809–1811), mit einer Einführung von Marco Segala*. Göttingen: Universitätsverlag Göttingen.

Zammito, John H. 2012. The Lenoir Thesis Revisited: Blumenbach and Kant. *Studies in History and Philosophy of Biological and Biomedical Sciences* 43 (1): 120–132.

———. 2017. *The Gestation of German Biology. Philosophy and Physiology from Stahl to Schelling*. Chicago: University of Chicago Press.

Person Index[1]

[1] Note: Page numbers followed by 'n' refer to notes.

M. Kisner, J. Noller (eds.), *The Concept of Drive in Classical German Philosophy*,
https://doi.org/10.1007/978-3-030-84160-7

Subject Index[1]

[1] Note: Page numbers followed by 'n' refer to notes.

M. Kisner, J. Noller (eds.), *The Concept of Drive in Classical German Philosophy*,
https://doi.org/10.1007/978-3-030-84160-7

The manufacturer's authorised representative in the EU is Springer Nature Customer Service Centre GmbH, Europaplatz 3, 69115 Heidelberg, Germany. If you have any concerns regarding our products, please contact ProductSafety@springernature.com

Printed and bound by CPI Group (UK) Ltd, Croydon, CR0 4YY
30/09/2026
02225704-0006